WHOLESOME
HARVEST

ALSO BY CAROL GELLES

The Complete Whole Grain Cookbook

WHOLESOME HARVEST

COOKING WITH
THE NEW FOUR FOOD GROUPS —
Grains, Beans, Fruits, and Vegetables

CAROL GELLES

Foreword by Neal D. Barnard, M.D.
President, Physicians Committee for Responsible Medicine

LITTLE, BROWN AND COMPANY
BOSTON • TORONTO • LONDON

First Edition

LIBRARY OF CONGRESS CATALOGING-IN-PUBLICATION DATA
Gelles, Carol.
    Wholesome harvest : cooking with the new four food groups :
grains, beans, fruits, and vegetables / Carol Gelles. — 1st ed.
        p.   cm.
    Includes index.
    ISBN 0-316-30735-1
    1. Vegetarian cookery.   2. Proteins in human nutrition.
I. Title.
TX837.G39   1992
641.5'636 — dc20   91-44476

10 9 8 7 6 5 4 3 2 1

DESIGNED BY JEANNE ABBOUD

MV-NY

Published simultaneously in Canada
by Little, Brown & Company (Canada) Limited

PRINTED IN THE UNITED STATES OF AMERICA

With love and appreciation for

Mom, Dad, and Sherry,
Randy, Paula, and Poppy,

who have all brought joy into my life

Contents

Foreword   ix

Author's Note   xi

Acknowledgments   xii

Introduction   xiii

The New Four Food Groups:
   Guidelines for a Healthier Lifestyle   1

Cooking Basics   43

The Recipes   65
   Soups   71
   Main Dishes   99
   Side Dishes   146
   Salads   180
   Breads and Spreads   212
   Breakfast Dishes   251

Bibliography   265

Index   267

Foreword

On April 8, 1991, the Physicians Committee for Responsible Medicine proposed a profound change in nutritional policies. Our aim was to bring federal nutrition guidelines into line with modern research showing the power of grains, beans, fruits, and vegetables.

In this book, Carol Gelles translated PCRM's New Four Food Groups proposal into practical information, from menus to cooking tips to delicious recipes and information on how to select — or substitute — recipe ingredients. Millions of people want to eat in a healthful way but are not sure how to go about it. This book makes it easy.

The menus and recipes presented here are delightful. They range from light to savory, and from familiar to exotic. And while they will please the palate, they also have effects that linger long after. Making these recipes part of your routine can cut cholesterol levels dramatically, even for many people who have been frustrated by less stringent diets, and will encourage weight reduction, because they are devised with an eye toward boosting complex carbohydrates and reducing fat. The good news is that weight reduction with the New Four Food Groups plan tends to be permanent. That means no more diets. With these recipes, you can also reduce the risks of a broad range of other health problems, including cancer, stroke, and diabetes.

At the same time, this plan provides the protein, calcium, vitamins, and minerals the body needs, including the natural chemicals contained in fruits and vegetables recently shown to boost the immune system and slow the aging process. As a doctor, I am most fond of this aspect of the New Four program. You hold in your hands a potent prescription. Many vegetables are rich in beta carotene, which strengthens the immune defenses against invading bacteria and even against cancer cells. Beta carotene and other building blocks of vegetables and fruits also help neutralize toxins that are part of the aging process. Grains, fruits, vegetables, and beans are packed with soluble fiber, which lowers cholesterol levels, and insoluble fiber, which helps prevent cancer and maintains a healthy digestive system. Collectively, these foods are very powerful medicine.

As you explore these recipes, you will find your tastes drifting further and further away from the high-fat fare on which so many of us were raised. That was certainly true for me, coming from a midwestern family in which doctors were outnumbered by cattle raisers. The pork chops and roast beef that were my culinary routine gave way to a new kind of eating that is not only endlessly varied, but also the foundation for building a healthy and strong body.

Neal D. Barnard, M.D.
President, Physicians Committee for Responsible Medicine

Author's Note

I grew up in the 'fifties, when good nutrition meant plenty of protein, and plenty of protein meant plenty of meat. At that time, it was not unusual for me to have three meat meals a day. Comfort food has always been meat or dairy for me. Meat loaf, chocolate pudding, scrambled eggs — these were foods that made me feel safe and secure.

Over the years, as I followed the scientific research related to nutrition, I started making moves toward a healthier eating profile. I added more grains to my diet, I ate meat less frequently, I cut back on my egg intake, but I felt incapable of ever giving up my greatest vice, high-cholesterol foods: buttered bread, or even better, buttered bread with cheese . . . cream in my coffee . . . prime meats, bacon, salami . . . need I go on?

All that changed like a clap of thunder. I woke up one day and realized that at forty-three I was only seventeen years younger than my mother was when she had her first heart attack. Seventeen years certainly sounds like a lot of time, but I was learning that it goes by in the blink of an eye. On the other hand, it is, hopefully, enough time to turn myself around — eat right, get active, lose weight — time enough to get healthy, and, maybe, time enough to avoid or at least delay the possibility of following in my mother's footsteps.

Now don't get me wrong, I'm no paragon of virtue — I'm a MOTT (a phrase borrowed from my friend Randy), a Most Of The Time. I eat those things that I enjoy and are good for me, most of the time. Occasionally I indulge in a juicy steak or a cracker with Brie — but that's okay, because I've found that most of the foods that I eat regularly are foods that I truly enjoy; I don't feel deprived eating these healthful foods. It's my hope that by sharing my recipes with you, I'll help you see that eating healthfully doesn't mean giving anything up. It's a means to good health and feeling good about yourself as well. *Bon appétit!*

Acknowledgments

This book could never have been completed, and probably never even started, without the help and support of many people.

Many thanks to my colleagues Pat Baird and Lorna Sass, who suffered with me through tastings of unusual ingredients and whose advice and encouragement were extremely valuable. And to Holly Garrison and Chris Koury, who were always available to answer food questions and to listen to various versions of copy.

Thanks to Jennifer Josephy, my editor, whose accessibility and willingness to consider and go with new ideas have made this project a pleasure; to Deborah Jacobs, my copyeditor, whose perfectionism provided just the finishing touches the manuscript needed; and to Judith Weber, my agent, who was generous with her advice and helped to give shape to a not totally formed concept.

I am grateful to the Physicians Committee for Responsible Medicine, headed by Dr. Neal Barnard, for the concept of the New Four Food Groups. And to those friends and neighbors (most notably Marie and Bob Riesel) who tasted and tasted and tasted one recipe after the next.

A special thanks to my family and friends, who were always there.

Introduction

The purpose of this book is twofold: to serve as a guide to the New Four Food Groups — grains, beans, fruits, and vegetables — and as a cookbook for people looking for interesting new meat-free dishes. For those seeking a healthier lifestyle who are not comfortable with vegetarian recipes, *Wholesome Harvest* offers recipes that are nutritious and delicious substitutes for and/or accompaniments to meats.

I've tried to keep in mind that many of us run on incredibly crowded schedules and that soaking and cooking beans for hours is not desirable or feasible. Whenever possible, I've called for already cooked beans in recipes; this allows you the choice of cooking dried beans from scratch or using canned beans. I think many brands of canned beans are extremely acceptable in terms of taste and texture, and the nutritional values are not all that different (except for sodium, which is quite high) from home-cooked. If using a can of beans allows you the time to prepare the recipes in this book, then I strongly encourage you to do it!

The recipes were not developed for any particular therapeutic diet, although I did try to create ones that are generally healthful. You will find that many of the recipes have a moderate amount of fat, which would be required by someone using regular (not nonstick) cookware. If you are on a low-fat or low-sodium diet, you can alter these recipes to suit your specific needs (see "Nutritional Strategies," pages 26–32, for instructions).

When browsing through the recipes, keep in mind that entrées and side dishes are interchangeable in most instances. With a simple change in portion size, most of these entrées can be served as side dishes and, conversely, the side dishes can easily be used as main dishes. Since many of the salads are also made of hearty, satisfying grains and/or beans, these can be converted into entrées by altering portion sizes as well.

I find that the best way to fit the New Four Food Groups into my life is to take a little time to plan ahead. Although many of these recipes are easy or quick to make, they still require that you have the ingredients on hand. So keep a good variety of fresh fruits and vegetables at home and have quick-cooking grains, as well as longer-cooking ones, in your pantry. Have cans of beans available or precook beans from scratch. Then, even when your life gets hectic, you'll be able to throw together a great, good-for-you meal.

1.

The New Four Food Groups:

GUIDELINES FOR A HEALTHIER LIFESTYLE

SINCE THE DAWN of civilization, humankind has pursued longevity and good health. It is only now, with advances in medicine and scientific knowledge, that real strides have been made in extending life expectancy. Unfortunately, most life-extension methods involve keeping people alive longer once they have become ill. Yes, medication and modern surgical techniques can extend people's lives — but is that the most we can expect?

Scientists have also given us clues to a healthier old age, but these require work and changes in the way we live in order to prevent illness. That dread word "exercise" keeps popping up, and then that other loaded word, "diet."

In the 'fifties the U.S. government came out with a revolutionary concept of "good eating": the four food groups. By eating the proper amounts from each of the four groups, Americans would be getting at least 80 percent of their daily requirement of vitamins and minerals and would be assured of good nutrition.

These "basic four" were:

1. Meat, poultry, fish, eggs, beans, and nuts
   2 or more servings daily
2. Milk and dairy products
   2 or more servings daily
3. Bread and cereal
   4 or more servings daily
4. Fruits and vegetables
   4 or more servings daily

Although the number of servings for the bread-cereal and fruits-vegetables groups were twice those of the meat and milk groups, the emphasis for good nutrition was truly on meat and milk. Further, while beans and nuts were considered part of the meat group, hardly a word was ever said about the possibility of eating beans instead of meat; the thought was almost un-American.

Now, more than thirty years later, scientists are finding that good health is actually better served by making the major part of your diet the bread-cereal and fruits-vegetables groups and de-emphasizing foods high in saturated fats, such as meat and cheese.

The concept of the New Four Food Groups — grains, beans, fruits, and vegetables — was presented in a report in April 1991 by the Physicians Committee for Responsible Medicine, a group of two thousand physicians. The report suggested that the American diet should shift away from animal products and high-fat foods and toward the New Four, which are high in fiber and complex carbohydrates. Dairy products and meat, therefore, should no longer be considered food groups but rather optional food items, and for maximum health benefits, all animal products should be eliminated from the diet.

According to the report, a person switching to the New Four plan will significantly reduce his or her likelihood of contracting many of the debilitating diseases that destroy our health and the quality of life as we age.

Some of the conditions for which the risk can be lowered or eliminated by the New Four are cancer and heart disease. Since breast cancer is associated with high fat intake, the lower fat in this program will reduce this risk factor. Colon cancer is also associated with a diet high in animal fat. Limiting the intake of animal products, as well as increasing the consumption of fiber in the diet, reduces the risks of this type of cancer. Strict vegetarianism has also been linked to decreased risk of lung and ovarian cancers.

There are many factors in the New Four plan that contribute to lowered risk of heart disease. Reducing fat and cholesterol intake (especially animal fats, which are so closely associated with saturated fat and cholesterol) and increasing fiber intake (especially soluble fiber from beans, oat bran, rice bran, and apples, which has been associated with lowering cholesterol) are significant methods of decreasing the risk of heart disease. Another factor is reduced obesity. The New Four program eliminates many of the foods that contribute to obesity, and with decreased obesity comes reduced risk of heart disease.

A strict vegetarian diet has also been associated with reduced risk of high blood pressure, diabetes, gallstones, kidney stones, osteoporosis, and asthma. Life without these complications is certainly more in line with our dreams of longevity.

How Do the New Four Differ from the Old Four Food Groups?

At first glance there seems to be an enormous difference between the food groups recommended in the 1950s and the New Four. But in fact, the U.S. government's emphasis has been shifting from meats and dairy to grains and cereals. This shift can be seen in its latest recommendations, represented by the Eating Right Pyramid. These recommendations make fruits and vegetables two separate groups — thus giving more emphasis to each one — and suggest reducing saturated fat to less than 10 percent of total calories. The five food groups as described in the Eating Right Pyramid are:

- *6 to 11 servings daily of bread and cereal,* instead of the 4 servings previously recommended
- *2 to 4 servings daily of fruit,* an increase from previous recommendations
- *3 to 5 servings daily of vegetables,* also an increase from previous recommendations
- *2 to 3 servings daily of meat*
- *2 to 3 servings daily of dairy products*

The New Four Food Groups plan (see pages 22–24 for allowances and serving sizes) takes the recommendation about reducing calories from saturated fat — which, in effect, means reducing the consumption of animal products — one step further and suggests that, ideally, for maximum health benefits, a completely vegan (no meats or meat by-products, including eggs and dairy products) diet is best.

Such a regimen may seem drastic and not necessarily acceptable to the population in general; it's more extreme than the diet many vegetarians follow, which includes eggs and dairy. However, since health is not an all-or-nothing proposition, you can derive health benefits by following a less strict food plan — which I call the Modified New Four Food Groups (see page 24).

Vegetarianism Is Not for Me!

The object of the New Four Food Groups plan is not to convert you to vegetarianism. The vegan diet is more like a guiding light. Any steps that you take in your current eating habits that go in the direction of that light will bring you health benefits. The more steps you can take, the larger the rewards you will reap.

Don't despair if you can't imagine yourself ever giving up meat or dairy. You don't have to! You can follow the modified New Four plan, which includes *up to* 3½ ounces of meat and optional dairy products each day (see "The Modified New Four Food Groups Eating Plan," page 24, and "Strategies for the Modified New Four," pages 29–31).

If even that seems like too much of a sacrifice, make whatever changes you can live with. When it comes to eating meat and dairy, the expression "less is more" is the brass ring to reach for. Small changes, such as eating smaller portions of meat (a four-ounce instead of an eight-ounce burger), switching from mayonnaise to mustard or from butter to jelly, cutting down from two eggs to one, eating cereal instead of eggs for breakfast, or having an occasional meatless meal, may be steps that you will find doable. If you're reading any of these suggestions and saying to yourself, "Okay, I can do that one," you will find that, eventually, all the little changes will mount up to big rewards.

❦

THE NEW FOUR AND GENERAL NUTRITIONAL PRINCIPLES

ALTHOUGH THE NEW FOUR eating plan may at first seem to you like a radical shift from your current diet, it is designed to meet all your nutritional needs. It provides plenty of all of the three major food substances that human beings require — protein, carbohydrates, and fats — and of fiber and vitamins. The plan also allows a sensible approach to limiting cholesterol and monitoring calories.

A PROTEIN PRIMER

By the fourth grade, every student is taught that proteins are the building blocks of the body, but what exactly does that mean?

Proteins are made up of twenty-two amino acids. Fourteen are produced in our bodies, but eight must be obtained from the foods we eat: phenylalanine, isoleucine, leucine, lycine, methionine, threonine, tryptophan, and valine. We call these eight the essential amino acids (EAAs). Amino acids are the materials used to form the structural parts of all cells, including antibodies, enzymes, and hormones. Since our bodies are in a constant state of cell renewal, we must provide them with sufficient materials for this function, and that means eating enough of the right foods to supply the necessary EAAs.

When a food contains all eight of the essential amino acids, in just the right proportion, that protein is considered a **complete protein.** These complete proteins are foods that the body uses efficiently, with little waste. They are found in meat, dairy products, and other foods of animal origin.

Foods that are of vegetable or plant origin also contain all eight

EAAs, but these are almost never in the proper ratio to one another to be used entirely by the body; they are considered **incomplete proteins.**

To understand the importance of the ratio of the EAAs, imagine that your body is a construction company. It has a certain blueprint that shows how the building (complete protein) must be constructed. For this building, each brick (amino acid) must be a certain size and shape to fit the blueprint.

When you present this company with a pile of bricks made of animal protein, the workers are able to build the house according to the blueprint without any problem, using all of the bricks without any left over.

The building would look like this:

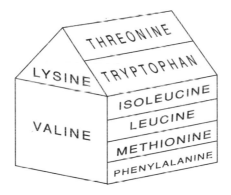

ANIMAL PROTEIN

The building made of the vegetable protein "bricks" that form rice would look like this:

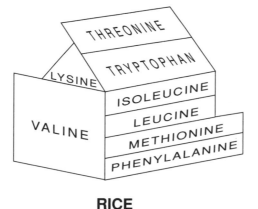

RICE

In order to use these vegetable bricks, the body finds the "limiting" amino acid (the one in the smallest amount — in this case, lysine) and trims the remaining amino acids so that the building will be in proportion to the amount of the limiting amino acid provided. The building made of the vegetable protein in rice, after the excess amino acids are whittled away, would look like this:

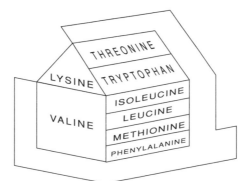

USABLE PROTEIN FROM RICE

The large outline represents the total amino acid content of rice, and the small house within the outline is the usable protein determined by the limiting amino acid.

But instead of wasting the excess amino acids from the rice, our hypothetical construction company, the body, has a few options: it can use them as a source of energy, convert them to fat for some future use, or combine them with another set of bricks, or protein, to build a perfectly proportioned complete protein. The sketches below illustrate how the EAAs that compose the protein in black beans provide an exact fit with those in rice. Black beans have a surplus of lysine, the limiting amino acid in rice, and a shortfall in methionine and cystine, which are overabundant in rice.

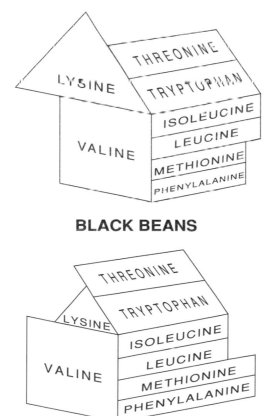

BLACK BEANS

RICE

The body mixes the extra lysine in the beans with the reduced lysine in the rice and treats the two like one source of protein. It does the same with cystine and methionine. Together the two sets of bricks act as **complementary proteins** and form the house below, with shared strengths indicated by cross-hatching:

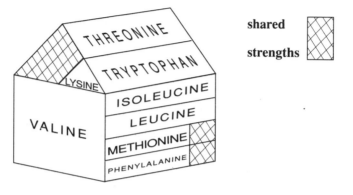

shared
strengths

BLACK BEANS AND RICE

How do the principles of complementary proteins fit in with the New Four Food Groups?

Two of the New Four Food Groups are grains and beans. As we saw in the preceding section, these two foods are excellent complementary proteins. By following the New Four plan, you are automatically making complementary protein matches. You don't have to eat complementary proteins at the same meal to derive benefits from this combination, but it is best to have them both within the same day. This is why it is suggested that you eat both grains and beans every day.

If you are following the Modified New Four Food Groups plan, the meat and dairy products that you eat all complement both grains and beans, so again, you don't have to worry about not having an adequate protein intake.

Isn't animal protein better than vegetable protein?

The obvious solution to getting enough high-quality protein seems to be to eat meats and not bother fussing with complementary proteins. On the other hand, animal products come with lots of undesirable traits along with the good ones. Let's look at some of the numbers:

Amount	Food	Sodium (mg)	Cholesterol (mg)	Saturated Fat (g)	Fiber (g)
4 oz.	lean beef	57	108	6.4	0
4 oz.	chicken (skinless)	84	96	1.1	0
4 oz.	flounder	119	77	0.4	0
4 oz.	kidney beans	3	0	0.1	9.6
⅔ cup	brown rice	0	0	0.2	1.9

Figures calculated using "The Food Processor II" (software), ESHA Research, Salem, Ore. All foods are cooked without added salt.

As you can see, the animal proteins are higher in sodium, cholesterol, and saturated fat than the vegetable sources, and lower in fiber.

In addition, since animals raised for food are fed hormones to help them grow faster and antibiotics to prevent disease, residual chemicals can remain in the meats we eat. It is true that vegetables are grown using pesticides, but animals eat large quantities of grain and therefore are exposed to pesticides as well.

Switching to a diet that includes some, or all, vegetable proteins will increase your intake of dietary fiber and reduce your intake of cholesterol and saturated fats. One word of warning to anyone who is considering a vegan (no meat, dairy, or egg) diet: vitamin B_{12} is not widely found in vegetables, so be sure to supplement your diet with this very important nutrient.

CARBOHYDRATES

Back in the dark ages, when I was growing up, carbohydrates were to dieters what garlic is to vampires. Fortunately, science has come around 180 degrees, and carbohydrates are now looked upon favorably by dieter and nondieter alike.

One of the first revelations to improve the carbohydrate image was that fiber is good for you. So good, in fact, that everyone began gobbling up anything that included the word "bran" on the label. Then, when soluble fiber became the cholesterol cure-all of the minute, oat stocks skyrocketed; suddenly you could find oat bran in everything from cereals and breads to cookies and waffles.

I'm not denying that bran is good for you but rather am emphasizing that carbohydrates have been good for you all along. They are a major source of energy for your body, and as such they are "protein sparers" — that is, they spare proteins from being used as an energy source. If your body doesn't get sufficient energy (calories) from the carbohydrates and fats you are eating, it will use the proteins you consume as a source of energy instead of using them as proteins. Imagine a fantastic engine that has run out of gas and can convert the oil into gas to keep running. That will work in the short term, but in the long term running a car without oil functioning as oil will burn out the motor entirely. Similarly, your body needs to use protein as protein, not as energy. This protein-sparing function of carbohydrates is especially important when you are reducing your protein intake.

In addition to their role as protein sparers, carbohydrates are necessary for fat metabolism and for maintenance of nerve tissue. They also serve as sources of "brain food" and of vitamins, minerals, and, of course, fiber.

But not all carbohydrates are created equal. Simple carbohydrates are simple in molecular structure and are not of much value to the body except as a source of pure energy. In the food world these are what sugars and other sweeteners are composed of. Complex carbohydrates are made up of many different molecules. In addition to energy, they provide the body with fiber, vitamins, and minerals. Fruits, vegetables, and starches are all complex carbohydrates, and they are the kind of carbohydrates that should be included in the diet; simple carbohydrates should be kept to a minimum.

Just as protein rarely comes in its pure form but is usually mixed with some level of fat, starches are not pure carbohydrate. Many starches are also composed, to varying degrees, of protein. Those with greater protein content (grain and grain products; starchy vegetables such as potatoes, yams, and peas; and legumes) are the basis for complementary proteins (see pages 6–10).

As for the dieter, the good news is that current research shows that carbohydrates, in addition to being relatively low in calories (ounce per ounce, pure carbohydrates have the same calories as pure protein and less than half the calories of fat), require extra energy to digest, therefore making them better for weight loss than protein or fat. So bring on the potatoes, bread, grains, and beans, and eat your way to better health and healthier weight loss.

FATS

The word "fat" strikes terror into most people's hearts (not to mention their arteries). Although too much of the "wrong" fats may, in fact, be unhealthful, some fat is an essential and beneficial part of your diet. On the other hand, the connection between cholesterol and saturated fat and heart disease has been clearly demonstrated, and high intakes of fat are also implicated in certain types of cancer, as well as less deadly conditions like gout, gallbladder problems, and obesity. Clearly, then, it's a question of how much and what kind of fats you consume.

Fat serves as a source of concentrated energy for the body. In addition, it is an essential part of cell membranes and aids in the transport of fat-soluble vitamins (A, D, E, and K). Fat deposits, which concern many people, have very important uses: they help hold the organs and nerves of the body in place and protect them from shock; they insulate the body and help maintain body temperature; and they act as a storage depot for energy reserves.

Fats are divided into three types, depending on their chemical structure: **saturated, monounsaturated,** and **polyunsaturated.**

Saturated fats have no molecular double bonds, typically are solid at room temperature, and are usually found in foods of animal origin or in tropical oils. These fats should be avoided as much as possible. They are the primary culprits in contributing to poor health.

Monounsaturated and polyunsaturated fats have one (mono) or more than one (poly) molecular double bond and are liquid at room temperature. They can be used in your food plan, but keep an eye on amounts, because too much of any fat is not good for you.

❧ COMPARISON OF CHARACTERISTICS OF DIFFERENT FATS ❧
(PER TABLESPOON)

Fat	Calories	Cholesterol (mg)	Total Fat (gm)	Saturated Fat (%)	Unsaturated Fat Mono (%)	Poly (%)
Butter	108	33	12	62	30	5
Canola oil	120	0	14	7	56	33
Coconut oil	120	0	14	87	6	2
Corn oil	120	0	14	13	24	59
Lard	116	12	13	39	45	11
Margarine	102	0	11	20	45	32
Margarine, diet	49	0	6	20	40	37
Olive oil	119	0	14	13	73	8
Peanut oil	119	0	14	17	46	32
Safflower oil	120	0	14	9	12	74
Vegetable shortening	113	0	11	25	45	27

Choosing a fat to cook with

As you can see from the table "Comparison of Characteristics of Different Fats," not all fats are the same. In following the New Four plan, your objective is to choose fats that have no cholesterol and that are low in saturated fats. It's clear that butter and lard should be eliminated from your cooking repertoire, because they both contain cholesterol and are high in saturated fats. Coconut oil should be cut out because of its high saturated-fat content.

Margarine is an ingredient I use, but not too often. It is not especially good for you; recent research indicates that health benefits derived from switching to margarine from butter are minor. I use margarine when a buttery flavor is desirable (to impart a sense of richness), as a base for cream sauces, and in some baked goods. If you are a vegan, be sure to read the labels carefully before choosing a margarine; many contain animal products.

Oils come in various nutritional profiles and flavors. Nutritionally, all oils are about 120 calories per tablespoon and are 100 percent fat. The oils that are best are those highest in monounsaturated and poly-unsaturated fat. The three best oils are canola, olive, and safflower. Each is slightly different from the other.

Canola, which has the lowest saturated-fat content of any fat, is definitely high on the health-benefits scale. It has a neutral, almost undetectable flavor. You can use it for all cooking and baking purposes, and you can also substitute it for olive oil, if you don't like the flavor of olive.

Olive oil, which I have called for in some of the recipes in this book because of its benefits as a flavoring agent as well as its properties as a fat, is distinguished by its "pressing." The first pressing of olives is labeled extra virgin. The oil produced from this pressing is the finest, best-tasting olive oil (and the most expensive, too); it is usually green or greenish. The second pressing is labeled virgin and is less flavorful and more yellow than extra virgin. The pressings after virgin are simply called olive oil and are usually deep yellow with little or no hint of green. The different pressings create a wide range of flavors. Finding the flavor that appeals to you is a matter of trial and error. My own preference is for the light, fruity flavor of extra-virgin Colavita oil. If you find the more flavorful olive oils objectionable, try olive oil from later pressings, which are less distinctive in flavor. Olive oils labeled light are those with the least olive oil flavor, but they still have the same grams of fat and number of calories as regular olive oil. A look at the table on page 13 shows olive oil to be the highest of all fats in monounsaturated fat, a fat that is currently being investigated for a possible role in preventing heart disease. Generally, olive oil is not suitable for sweet items such as muffins, pancakes, waffles, or desserts.

Safflower oil, which has a mild flavor, is extremely high in polyunsaturated fat and very low in saturated fat. Try it to see if it appeals to you. You can use it in any recipe calling for oil, including olive oil.

Oils that are simply designated vegetable oil, such as Wesson and Crisco, are a mixture of different vegetable oils. These oils, as well as any others of plant origin (such as corn or peanut), are nutritious but not as good as canola, olive, or safflower, which score either high in unsaturated fats or low in saturated fats.

In addition to the oils mentioned above are sesame, chili, and nut (almond, walnut, and hazelnut) oils, which are specialty oils used in small amounts as flavoring agents. They are all from the vegetable kingdom and are fine nutritionally. Sesame oil is strong in flavor and frequently used in Oriental — especially Szechuan — dishes. It is normally stirred in at the end of cooking, rather than for sautéing. Chili oil is usually made from sesame oil and has the added benefit (if you like hot food) of being *very* spicy. It will add fire to anything. Use it sparingly, to your taste. If you don't like spicy foods, substitute equal parts of sesame oil for chili oil.

CHOLESTEROL

Like fat, cholesterol is essential to your body. Its purposes include the production of hormones and cell membranes and the conversion of sunlight into vitamin D. Cholesterol is a waxy substance produced in the liver and also obtained from the foods we eat. However, since we produce enough cholesterol in our bodies to fill our needs, it is not necessary to seek out additional cholesterol in food.

With cholesterol, too much of a good thing can lead to unhealthy outcomes. Contrary to popular conceptions, there are no "good" and "bad" cholesterols. The risk of heart disease as related to cholesterol depends on which of the two different carriers transports the cholesterol through the blood. Cholesterol attached to high-density lipoproteins (HDL) is considered beneficial because it is brought to the liver to be used or eliminated from the body. Cholesterol attached to low-density lipoproteins (LDL) is associated with high risk of heart disease because the LDLs tend to deposit it on artery walls. As the cholesterol builds up, the artery openings become smaller and smaller. This clogging of the arteries is known as atherosclerosis and can result in heart attack, stroke, and even death.

Cholesterol is not the only culprit in clogged arteries. Saturated fats play a large role in atherosclerosis. As you increase your intake of saturated fats, the level of LDLs also increases. Monounsaturated fats, and to a lesser degree polyunsaturated fats, have been found to lower

LDLs. The most current findings on risk of heart disease and cholesterol indicate that the ratio of LDLs to HDLs may be even more important than your cholesterol count itself.

How much cholesterol and fat am I allowed?

Ideally, you should avoid cholesterol completely. By following the New Four Food Groups plan this is easy, since cholesterol is found only in animal products and the New Four eliminates them from the diet. The Modified New Four program, however, does allow meat and dairy products. It is strongly suggested that you keep animal fats to a minimum, sticking to cuts of meat that are very lean, poultry without skin, fish and seafood, and low- or no-fat dairy products.

According to the guidelines of the New Four plan, total fat intake should not exceed 20 percent of your total calorie intake. That figure is the same for the Modified New Four. The government guidelines are slightly more generous and suggest 30 percent fat from calories. I'm sure these figures are familiar, but few of us actually calculate the percentage of fat we consume each day. How do you do it? Here's the formula:

Step 1: number of grams fat × 9 (the number of calories per gram of fat) = number of calories from fat

$$\text{Step 2: } \frac{\text{number of calories from fat}}{\text{total caloric intake}} = \text{percentage of fat}$$

An example will make this clearer: If yesterday I consumed 1,800 calories, which included 45 grams of fat (these figures can be found on nutritional labels), I would first translate the grams of fat into calories:

$$45 \text{ grams fat} \times 9 = 405 \text{ calories from fat}$$

Then I would divide that figure by my total caloric intake:

$$\frac{405}{1,800} = 22.5 \text{ percent fat}$$

According to these figures, I would have eaten 2.5 percent more fat than the New Four program suggests (to meet the 20 percent, I should have consumed only 40 grams of fat), but 7.5 percent less fat than the government recommends.

This probably still doesn't mean a whole lot to you, so here are some examples of 45 grams of fat:

    3 tablespoons plus 1 teaspoon oil
    4 tablespoons (¼ cup) margarine or butter
    4 tablespoons (¼ cup) mayonnaise
    6½ tablespoons Italian dressing
    ½ cup heavy cream
    1 cup sour cream
    4½ ounces cream cheese
    5½ cups whole milk
    15 cups 1% milk
    ⅓ cup peanut butter
    1 eight-ounce cooked hamburger patty
    8 large eggs
    14 slices bacon
    5 ounces hard or semisoft cheese
    1 Big Mac and 1 regular fries
    1 Whopper with cheese
    1¼ cups Häagen-Dazs chocolate ice cream
    4½ ounces potato chips

A reminder: just because you're eating something from the plant world doesn't mean it's automatically fat-free. For instance:

½ medium avocado	15.0 grams fat
1 tablespoon peanut butter	8.0 grams fat
10 large black olives	6.5 grams fat
10 large green olives	4.9 grams fat
4 ounces tofu	4.8 grams fat

Depending upon how you "spend" your grams of fat, 45 grams can seem like a lot or a little. If you eat a Whopper with cheese and use up all the fat for the day in that one sandwich, it will seem as though your fat allowance is pretty meager. On the other hand, if you are omitting meat and using oil only for cooking and salad dressings, you shouldn't feel too crimped at all by the 45 grams suggested.

However, this benchmark of 45 grams will not represent the upper limit of fat consumption for everyone. Remember, the percentage of fat is based on your total caloric intake. If you eat 2,000 calories, then the 45 grams of fat would be 20 percent of your caloric intake. If you eat only 1,200 calories, 27 grams of fat would be 20 percent of your total caloric intake. So, be aware that if your total caloric intake is lower or higher than the one recommended for your weight, you should also adjust the amount of fat proportionally.

If you think that counting calories and doing the other calculations to determine 20 percent of your total intake are too complex for you, here are some rough estimates of the amount of fat constituting 20 percent of the allotted calories for different weights. Figures are for a lightly active person, thirty years old. If you are a couch potato and/ or over thirty, your calorie and therefore fat allowances will be lower than those given; if you have an active lifestyle or are under thirty, your allowances will be higher.

	Male		Female	
Ideal Weight	Total Calories	Fat (g)	Total Calories	Fat (g)
100	2,225	49	1,792	40
125	2,469	55	1,991	44
150	2,714	60	2,191	48
175	2,957	66	2,389	53
200	3,203	71	2,590	57
225	3,448	76	2,741	61
250	3,691	82	2,987	66

These figures assume that your daily caloric intake is average for that weight. You should strive to stay below these calorie totals for better health.

How much fat is there in the recipes in this book?

Most recipes have 8 or fewer grams of fat per serving, and many of these have far fewer. Each recipe is followed by a list detailing its nutritional makeup, including the total fat, saturated fat, and cholesterol it contains.

But these estimates for fat are probably higher than you will actually be getting, for two reasons. First, I included in each analysis all items labeled optional in the recipes, such as Parmesan. Second, I chose the highest-fat form of ingredients and the largest-size portion. For example, if a recipe called for one cup of milk, and it made four to six servings, I calculated the nutritional value using whole milk and serving four, which yielded the highest possible amount of fat for that recipe. As you can see, the fat content drops considerably with lower-fat milk and smaller portions.

Serving 4 using whole milk	2.0 grams of fat
Serving 4 using low-fat (1%) milk	0.75 grams of fat
Serving 4 using skim milk	0
Serving 6 using whole milk	1.3 grams of fat
Serving 6 using low-fat (1%) milk	0.5 grams of fat
Serving 6 using skim milk	0

In addition to using the lower-fat versions of the ingredients called for in recipes, you can further reduce the amount of fat in your foods in the following ways:

- Use nonstick cookware. For sautéing, you can reduce the amount of fat in recipes by one half to two thirds without a significant loss of flavor or change of texture.

- Use no-stick cooking spray instead of fats called for in sautéing. Unfortunately, some sacrifice of flavor and texture will occur.

CALORIES

The New Four Food Groups plan is not about calorie counting. If you eat according to the suggested allowances of food groups — keeping your fat intake below 20 percent of total intake — exact calories don't matter. If, however, you are finding that your weight is not going in the direction you desire, you may want to look at your caloric intake.

The calorie requirement is the number of calories a person needs to eat to maintain his or her desired weight. The number of calories you should eat is determined by a number of factors, including your weight, sex, age, and activity level.

For example (according to the National Research Council's *Recommended Dietary Allowances,* 1991):

For a twenty-three-year-old male couch potato who weighs 154 pounds, the calorie requirement is 2,406 calories per day; a man who is the same age and size but is very active would need 3,938 calories.

For a twenty-three-year-old female couch potato weighing 128 pounds, the calorie requirement is 1,856 calories per day. A similar female who is very active will need 3,038 calories.

Exactly how many calories you need will vary even from day to day, depending on your activity level. If you want to count your calories and need a guideline, consult the chart on page 18 for an approximate number.

FIBER

Fiber, which used to be referred to as roughage, is the indigestible part of plants. Its previous claim to fame was that it kept you "regular."

In recent studies, water-soluble fiber (fiber that dissolves in water) — found in beans, oat bran, rice bran, corn bran, and barley and apple fiber — was seen to have a correlation to lowered cholesterol. Whether soluble fiber directly affects cholesterol levels may be in doubt, but the fact that it contributes to general good health is not in dispute at all. Nonsoluble fiber, found in wheat bran and carrots and other vegetables, has not been linked to improved cholesterol levels but is valuable for many other aspects of good health (including that old standby, regularity).

If you are following the New Four plan, you are automatically getting plenty of fiber in your diet.

VITAMINS, MINERALS, AND TRACE ELEMENTS

If you are following the Modified New Four plan, getting all of the vitamins and minerals you need should not be much harder than for anyone who is also eating meat. In fact, since the emphasis of your diet is grains, beans, fruits, and vegetables, you are probably getting even more of most vitamins, minerals, and fiber than people who are following the guidelines of the old four food groups.

A little extra vigilance may be needed to fulfill the requirements for a few nutrients.

Calcium

The current recommended dietary allowance (RDA) for calcium is 1,200 milligrams daily for people eleven to twenty-four years old, and 800 milligrams daily for anyone older. If you are following the Modified program and you previously were able to meet the daily requirement for calcium, then you should find no difference now. If, on the other hand, you never ate enough calcium-rich foods, you probably won't be getting enough now either. Clearly, your best sources for calcium are dairy products, but what happens to people following the New or Modified Four plan, who eat no or very few dairy products?

The "Guide to Healthy Eating," a publication of the Physicians Committee for Responsible Medicine (the organization that originally proposed the New Four Food Groups), suggests that a diet lower in protein (which a vegan diet usually is) results in less calcium being eliminated in the urine. Therefore, if you are a vegan, you may need less calcium in the first place. Since this connection has not been proven conclusively, vegans are still best advised to try to obtain the full RDA of calcium. Doing so should not be hard, since there are many vegetarian foods with high calcium content, such as dark green vegetables (especially broccoli, kale, and collards), legumes, and tofu. Also, many products are coming out on the market that are calcium-fortified — most notably orange juice — and they can certainly help you avoid any calcium shortfalls.

Iron

Even if you are a red-meat eater, it's difficult to satisfy the requirements for iron, and it's especially hard if you are a menstruating woman. The RDA is 10 milligrams daily for males and 15 milligrams daily for females (and 30 milligrams if you're pregnant). There are many good sources of iron in a vegetarian diet, particularly dried fruits and legumes. Fortified cereals are an easily accessible source of iron. If you think you are not getting enough iron in your diet, consult your physician about a supplement.

Although vitamins D and B_{12} are not problematic to meat eaters, they can be difficult for vegans to obtain in the recommended daily amounts.

Vitamin D

Milk is commonly fortified with vitamin D, so if you're following the Modified plan, there should be no problem meeting the RDA for this nutrient. If you don't eat dairy products, you still have access to the best source of vitamin D around — the sun. Try to spend some time each day outside in the sun, leaving parts of your skin exposed, since sunlight is converted to this vitamin by way of the skin. This will not only boost your vitamin intake, it should also boost your spirits. (Don't, however, forget to wear sunblock.)

Vitamin B_{12}

The only vitamin not available in the vegetable kingdom is B_{12}. If you are following the vegan form of the New Four program, you may want to consider a vitamin supplement for this nutrient. Another source is fortified cereals, some of which contain your daily requirement for B_{12}. Check the label on the box to find out.

A final word on vitamins and minerals: variety is more than just the spice of life, it is a key to good health as well. Be sure to eat from the four food groups and indulge in as many different foods as possible each day.

~❧~

THE NEW FOUR FOOD GROUPS EATING PLAN

THE NEW FOUR PROGRAM is a simple guideline to healthy eating. It suggests how much of each of the four foods you should include in your daily menus. How you distribute these foods throughout the day is entirely up to you. If your lifestyle allows for only three meals a day, try to include the total number of servings among your three meals. If you like to "graze," eat a little of the recommended servings at each snack and try to meet the totals by the end of the day.

The following are important considerations for planning your New Four menus:

- The number of servings suggested is just a starting point. You can have more, but you shouldn't have less than the suggested amounts.

- Be sure to choose at least one fruit daily that is high in vitamin C (citrus fruits, melon, or strawberries).

- Make at least one selection daily from the dark green, leafy vegetables (bok choy, spinach, beet greens, collard greens, broccoli).

- Choose one dark yellow or orange vegetable or fruit daily (carrots, winter squash, sweet potatoes, cantaloupe).

- Remember: peanut butter and soy products are high in fat; use them judiciously.

- Keep your fat intake below 20 percent of your total calorie intake.

The serving allowances for the New Four plan, as established by the Physicians Committee for Responsible Medicine, are as follows:

Grains: 5 servings

> *serving size:* ½ cup cooked hot cereal
> 1 ounce ready-to-eat cereal
> 1 slice bread
> ½ cup cooked grain
> ½ cup cooked pasta

Grains, especially whole grains, are rich in complex carbohydrates and provide you with lots of fiber, B vitamins, minerals, and amino acids.

Grains are filling and satisfying. Use them freely as a base for main dishes and salads.

Beans: 3 servings

> *serving size:* ½ cup cooked beans
> 4 ounces tofu or tempeh
> 8 ounces soy milk
> 2 tablespoons peanut butter

Beans are a great source of complex carbohydrates, fiber, B vitamins, and amino acids. Three servings a day seems like quite a lot, but beans are so versatile that you can toss some into almost anything that you are preparing, whether it's salads, sauces, soups, or stews. Substituting soy milk for dairy milk is an easy way to fulfill part of your bean requirement.

Fruits: 3 servings

> *serving size:* 1 medium piece of fruit
> ½ cup fruit juice
> ½ cup cooked fruit

Fruits are rich in vitamins, minerals, fiber, and complex carbohydrates and are virtually fat-free.

Although there are lots of wonderful desserts you can make with fruit, you won't find any here. The reason is simple — the best possible way to eat fruit is raw! Raw fruits taste delicious and have all their nutrients intact. So eat fresh fruits as a dessert or a snack, or use them chopped in your salads, and save dessert recipes for special occasions.

Vegetables: 3 to 4 servings

> *serving size:* 1 cup raw vegetables
> ½ cup cooked vegetables

Vegetables are high in vitamins, minerals, and fiber. Most are fat-free, except olives.

In addition to being good for you, vegetables add a fresh taste and excellent texture to cooked food. Like fruits, they are usually at their best eaten raw (with the exceptions of eggplant, potatoes, winter squash, and the like). Go wild with your vegetable combinations for salads, stir-frys, soups, and anything you can think of. You don't have to limit yourself to four servings, but be sure to eat *at least* three; remember, the more vegetables the merrier!

A Note for Vegans

If you are a vegan, you're automatically following most of the recommendations of the New Four program. The only modification you may need to consider is controlling the percentage of your total calories that come from fat. If you're over 20 percent, you should try cutting down.

THE MODIFIED NEW FOUR FOOD GROUPS EATING PLAN

The Modified program includes all the suggestions for the basic plan but in addition allows *up to* 3½ ounces of animal products — beef, lamb, pork, chicken, fish, eggs, and hard or semisoft cheeses — daily. Because such cheeses (American, Swiss, cheddar, and Brie) are extremely high in fat, they should be eaten only occasionally, even at the prescribed daily amount. Other dairy products can be incorporated into the diet at your discretion, *over and above* the 3½-ounce ceiling, but try to use low- or no-fat versions.

WHAT'S FORBIDDEN?

Occasionally, you will crave dessert or other items not listed in the food groups. What should you do? Because the New Four plan is a way of life, not a diet you go on until you reach a specified goal, *there are no foods forbidden for the rest of your life.* The idea is to keep high-fat, low-fiber, and sugary foods to a minimum, but you don't have to eliminate them completely or forever.

Every day you'll be confronted with temptations; sometimes it will be a sinful and delicious dessert, other times an array of high-fat snacks at a party where you'll feel like joining in with the crowd. When these situations arise, ask yourself, Which is more important, my health or a momentary desire? If you're human, the momentary desire will sometimes win. The important point is to try to keep your goals in mind all the time and then use your conscience as an honest scale for each individual situation. If you usually don't eat rich foods, then indulging once in a while is okay. Just don't let those occasions become "usually."

The Sugar Question

Since, as I state above, there are no foods that are taboo on the New Four plan, you don't have to give up sugar and sweets, but you should keep them to a minimum. Try limiting sugary desserts to one a day or fewer.

I believe that some people are addicted to sugar (and I'm one of them, so I know whereof I speak). For us, the best way to conquer this problem is to treat sugar like any other addiction and just avoid it completely. The first few days may be excruciatingly difficult (as they are when you cold turkey any addiction), but then the cravings quiet down bit by bit.

The best approach is:

- Don't eat any prepared foods that have sugar or sugar products — corn syrup, honey, molasses, maple syrup, dextrose, fructose, maltose (any product that ends in "ose" is a sugar) as the first, second, or third ingredient on a product label. The government requires manufacturers to list the ingredients in descending order according to amount. Therefore, if sugar is the first ingredient, there is more sugar than anything else in that product.

- For desserts you prepare at home, use only recipes that have less than 1 teaspoon of sugar per serving. A teaspoon or even a tablespoon of sugar in a stew, sauce, or salad dressing is not going to undermine your efforts, assuming that the recipe serves at least four.

Believe it or not, the less often you eat sugary foods, the less often you will crave them. In fact, if you've given up sugar for a month or two and then decide to have something sweet, you may be surprised to find that it tastes *too* sweet. I've also found that when I eliminate meat completely, my craving for sweets diminishes.

Giving up sugar can be a difficult undertaking, but it's truly worth the effort.

HOW WILL THIS DIET AFFECT MY WEIGHT?

If you are switching to the New Four Food Groups, have eliminated all animal products completely, are keeping your fat intake below 20 percent of caloric intake, and are at your ideal weight, you will probably maintain your current weight. If you are overweight, you will probably lose weight without even trying.

Your weight should also be stable or reduced should you follow the

Modified New Four Food Groups plan and not choose high-fat meats or dairy products.

If you have a weight problem, reducing your fat intake will certainly help, but you should also keep your intake to minimum suggested servings and avoid foods with too much sugar in them.

NUTRITIONAL STRATEGIES

The New Four Food Groups plan, as you've already read, is a sure road to better health, whether you are choosing to become a vegan or to follow the Modified, low-meat regimen. For those of you who must watch your fat, cholesterol, sodium, or carbohydrate intake, the following pages outline methods of using this book and tailoring recipes and menus to be consistent with your needs.

Controlling Fat

Some people keep track of their fat intake for specific therapeutic reasons. Counting grams of fat is one popular method of weight loss, with a limit of 30 grams of fat a day a common ceiling. Those trying to control heart disease also limit their total fat intake (and they often follow programs that severely restrict protein intake as well). These people are on diets such as Pritikin or Dr. Ornish's, and counting fats and limiting protein are the most important dietary considerations.

If you are considering keeping track of your fat intake but you're not quite sure how to get started, the easiest way is to purchase a complete food counter. These guides, available in paperback in most bookstores, provide the fat content of most raw foods and many prepared ones too. If you use canned or other prepared foods in your cooking, you can find the fat content of those foods on their nutritional labels. For raw ingredients you will have to refer to the food-counting book.

Reducing Fat and Cholesterol in Recipes

What if a recipe, like the ones in this book, tells you how many grams of fat it contains per serving, but you would like to have fewer? The best way to reduce fat in the dishes you make is to invest in some good nonstick cookware, then substitute 1 teaspoon of fat for each tablespoon called for in a recipe. Why reduce the fat by only 66 percent — why not use no fat at all, or sprays such as Pam? Because, although you can reduce the fat content even further, fat adds more to a recipe than just calories. It also adds flavor, and it favorably changes the texture of foods when used for sautéing. I find that substituting 1 teaspoon for each tablespoon of fat leaves enough fat in the recipe to guarantee the flavor qualities I intended. In addition, there is some room, not to mention need, for fat in your diet.

You can also reduce fat and cholesterol in recipes by substituting low-fat products. This is especially easy when it comes to dairy products. Use 1 percent or skim (nonfat) milk instead of whole milk. Ricotta, mozzarella, and cottage cheese, as well as yogurt, all come in low-fat and no-fat styles. Other cheeses are also available in low-fat varieties. Even very high-fat foods, such as sour cream, cream cheese, mayonnaise, and margarine, all come in lower-fat versions. You can use yogurt as a substitute for sour cream and/or mayonnaise in most recipes.

Don't be deceived by "light" olive oil, however. It is *not* a lower-fat version of olive oil — it has lighter flavor but packs the same fat and calories as its "heavier" relatives.

Watching Sodium

The sodium content is included in the list of nutritional information following each recipe. I have included specific amounts of salt in the recipes for two reasons: to guide people who do use salt in their cooking, and to make clear the basis for the sodium count given in the nutrition analysis. Readers are encouraged to alter the amount of salt they use in these recipes according to their own tastes.

If you are trying to control the sodium content in your diet, you can choose those recipes with little or no sodium and/or omit any salt called for in a recipe. (My only caution concerns yeast breads: if you've left out the salt, you'll need to check your dough frequently to see whether it's doubled in bulk, since salt increases the rising time.)

It is better to skip certain recipes — most notably Oriental recipes, which rely heavily on soy sauce, an ingredient extremely high in sodium — rather than try to modify them. Even though soy sauce is available in reduced-sodium form, using it in a recipe calling for large amounts may still mean too high a sodium count for some.

A number of other ingredients, in addition to salt and soy sauce, are very high in sodium. One is prepared broth. Vegetable broth is available only in dried form. It comes either as a powder or a cube or in a packet, all of which are stirred into water to make broth. The dried vegetable broths available in the supermarket are all very high in sodium. There are some available in health food or natural food stores that are salt-free. You can make vegetable broth at home, and you will have a product that is not only salt-free but far superior in taste to anything currently available on the market. (For a great recipe for homemade broth, and more information about bouillon, see "Broth: The Flavor Enhancer," pages 58–60.) If you are on the Modified plan and are using canned chicken or beef broth, these are available in low-sodium form.

Unless it's otherwise stated, you can assume that anything that comes from a can has salt added. Some canned products are called for in the recipes. If you choose to use canned beans, which contain 400 to 600 milligrams of sodium per ½ cup, rinse them before adding them to the recipe to lower the sodium count. (Of course, for an even lower sodium count, you should cook your own beans from scratch.) Canned tomatoes are also sources of hidden sodium. I frequently call for canned whole peeled tomatoes in the recipes. You can use fresh tomatoes instead (about two large tomatoes, peeled, for one 15-ounce can). Many tomato products are available with reduced sodium, and you can use these as direct substitutes for the regular brands.

Cheese and other dairy products are high in sodium. Many of the recipes that use cheese (especially Parmesan) list it as an optional ingredient, so omit it if you're watching your sodium intake.

How much sodium is too much? Opinions have changed over the years. In 1980, for example, the RDA was 1,100 to 3,300 milligrams, then considered a "safe and adequate daily intake." The 1989 RDAs put the estimated minimum for healthy persons at only 500 milligrams daily, and today many researchers set the maximum at 3,000 milligrams, with the National Research Council recently suggesting 2,400 milligrams as a daily limit.

Here's a list of sodium levels in some products that you should watch out for:

Food	Sodium (mg)
1 teaspoon imitation bacon bits	230
1 teaspoon baking powder	300
1 cup canned chicken broth	800*
1 teaspoon lightly salted butter	35
1 ounce (unless otherwise specified) cheese:*†	
American	445
blue	360
cheddar	175
cottage (½ cup)	425
cream	90
feta	360
Monterey Jack	175
mozzarella	190

Food	Sodium (mg)
Parmesan	400
ricotta	60
Swiss	80
1 large egg	57
1 teaspoon garlic salt	1,850
1 ounce canned ham	340*
1 tablespoon ketchup	170*
1 tablespoon margarine	110
1 tablespoon mayonnaise	80*
1 cup milk	120
1 tablespoon peanut butter	90*
1 tablespoon bottled salad dressing	200*
1 tablespoon salsa	100*
1 teaspoon salt	2,713
1 teaspoon soy sauce	330
½ cup prepared spaghetti sauce	550*
½ cup canned whole tomatoes	220*
6-ounce can tomato paste	100*
½ cup prepared tomato sauce	600*

*Exact amount varies widely from brand to brand. The amount listed here is the average figure.
†Reduced-fat cheeses tend to be much higher in sodium than high-fat cheeses; check your package for figures.
List compiled using The Complete Book of Food Counts, *Corrine T. Netzer (Dell, 1988), and* "The Food Processor II" *(software), ESHA Research, Salem, Ore.*

My final thought on monitoring your sodium intake is: read labels whenever possible. You never know where sodium may be lurking.

Carbohydrate Counting

If you are looking for recipes low in carbohydrates, this is probably not the right book for you, since the foundation of the New Four Food Groups plan is complex carbohydrates.

STRATEGIES FOR THE MODIFIED NEW FOUR

If you've never tried to cut back on your consumption of meat and dairy products, following the Modified New Four program may seem difficult at first. But with a little practice and planning, you'll find very few sacrifices are required to follow this lifestyle.

Reducing your intake of animal products (and this includes not only all meats but also eggs and high-fat cheeses) to a maximum of 3½ ounces per day means that you'll usually have to limit your intake of such foods to once a day. That includes not just lunch and dinner but breakfast, too. If you're having two eggs for breakfast, they count toward your total meat intake — in fact, two large eggs (not even extra-large or jumbo) equal 3 ounces of animal protein. To stay under the 3½-ounce ceiling, you could have one egg for breakfast (1½ ounces meat), a salad for lunch, and pasta with marinara sauce topped with 1½ ounces of grated Parmesan to finish the day.

You'll also need to start counting hidden animal proteins, such as the eggs used in items like quiche, French toast, muffins, and pancakes.

Strategy 1. Plan your day in advance.

Decide at the start of each day which meal, or meals, will contain your animal product allowance and then plan the other meals accordingly.

For example, if you know you're going to a deli for lunch, have cereal with low- or no-fat milk for breakfast and use your meat on a sandwich at lunch. Then plan a meatless dinner. Or, if you're going to a barbecue for dinner, plan a meatless breakfast and lunch in anticipation of a grilled dinner. Don't go to the barbecue hoping that the hosts will offer a satisfying vegetarian alternative to the chicken or steak (unless you've called ahead and asked specifically for it — which is another good method of planning ahead).

Strategy 2. Keep a well-stocked larder.

Be sure to have plenty of supplies in your pantry and freezer so that you can easily put together delicious meatless meals (especially important for last-minute preparations).

Have pasta and sauces (store-bought or homemade) on hand. Homemade sauces can be made in large batches and then frozen in smaller portions.

Canned beans can easily be made into salads or main dishes with the additions of a few vegetables and some dressing.

Quick-cooking grains (such as white rice, couscous, bulgur, and quinoa) are also handy for that last-minute I'm-so-hungry-I-can-hardly-wait-for-dinner emergency. Couscous cooks in five minutes — faster than you can broil a burger.

Strategy 3. Make planned leftovers.

Cook large batches of soups, sauces, and stews. These foods freeze extremely well and can become instant meals in the near future. Similarly, extra home-baked breads can be sliced, then frozen for future use.

If you prepare beans from scratch or are cooking a batch of grains that take a long time to cook, make a double batch and use them in two different recipes: one for the same day, and one for later.

Strategy 4. *Make changes you can live with.*

You don't have to break all your habits at once. Start by cutting back on butter and on meats and cheeses that are highest in fat. Try cooking with vegetable oils instead of butter, and switch to lower-fat products, such as leaner meats and 2 percent milk instead of whole milk (eventually you may work your way down to skim milk).

Experiment with different recipes to discover meatless meals that you find satisfying. Then try to have one meatless day a week, and gradually increase the number of meatless days per week as you are able.

Strategy 5. *Cut back on all fats in general.*

The objective of the Modified and New Four plans is not to turn everyone in the world into a vegetarian but to make a healthier you!

One effective way to do this is to reduce the fat — especially animal fat — in your diet. Become an observer of your habits; cut out fats wherever you can. Use less, and eventually no, butter on your bread. Use less dressing on your salads. Use less fat in your sautéing. Use leaner cuts of meat. Every time you notice yourself using fats, ask yourself, "Can I cut down here?"

Since the Modified New Four Food Groups program is a way of life, not a temporary diet, there's no need to rush yourself to the point of discomfort. Every baby step is important, and the total of lots of small changes will ultimately be a big change toward a healthier lifestyle.

STRATEGIES FOR THE LACTO-OVO VEGETARIAN

If you're a lacto-ovo vegetarian — someone who eats eggs and dairy products but no other animal products — you probably feel that you're already pretty much in step with the Modified New Four. In some ways, you will find the changes involved to be greater than expected. The Modified and New Four plans stress the needs to reduce, as much as possible, all animal fats and even to cut back on vegetable oils.

Strategy 1. *Reduce your intake of eggs and high-fat cheeses.*

Many lacto-ovo vegetarians depend heavily on eggs and cheese as protein sources. The Modified New Four program makes dairy products optional, but favors eliminating them completely or at least limiting them. All cheeses that are high in fat should be regarded as meat and limited to 3½ ounces per day maximum. Eggs also count as meat and should be eaten sparingly. One large egg is almost half a day's meat allowance.

Strategy 2. Make butter an occasional food.

Switch from butter to vegetable-oil-based margarine whenever possible. The main emphasis of the New Four program is to reduce consumption of animal fat, and while butter is not 100 percent fat — it also contains milk solids — all of its fat *is* animal fat.

Instead of using butter on your breads, try some of the spreads in the chapter "Breads and Spreads" (such as Sesame Spread, page 244, or Hummos, page 247) or simply switch to jelly, apple butter, peanut butter (in moderation, since it is high in fat), or no-sugar-added fruit preserves.

Strategy 3. Use no- or low-fat dairy products.

If you don't already use them, switch to low-fat — or, even better — no-fat versions of such products as yogurt, cottage cheese, ricotta, and milk.

Strategy 4. Cut back on all fats in general.

This objective is also part of the Modified New Four Food Groups plan. Become an observer of your habits and cut out fats wherever you can. Use less, and eventually no, butter on your bread. Use less dressing on your salads. Use less fat in your sautéing. Substitute yogurt for all or part of the mayonnaise in a salad. Every time you notice yourself using fats, ask yourself, "Can I cut down here?"

❧

WHAT'S FOR BREAKFAST, LUNCH, AND DINNER?

IF THE ANSWER IS BACON and eggs, a tuna salad sandwich, and broiled chicken, figuring out how to follow the New Four plan may present a great challenge. Let's take it one meal at a time.

BREAKFAST

If you are keeping dairy (low-fat, preferably) and eggs in your diet, either as part of the Modified plan or a lacto-ovo vegetarian diet, breakfast can be as normal as cold or hot cereal with milk. Toast and muffins are fine, but try to ascertain whether the bread you're eating contains eggs so that you can count them in your 3½-ounce animal product maximum. Of course, low-fat breads and muffins are best. What to put on your bread is also a good question. Butter is not a good choice, and neither is cream cheese or hard cheese, since all of these are very high in animal fat. Instead, try fruit preserves, apple butter, peanut butter (occasionally — it too is a high-fat food), low-fat cottage cheese, some of the spreads included in this book, or (gasp!) nothing at all.

Waffles, pancakes, and French toast are acceptable breakfast choices, but remember that they contain eggs and fat, so you're better off having them infrequently, as part of a weekend breakfast or brunch. Yogurt or an occasional egg are other breakfast possibilities.

If as a vegan you eliminate dairy and eggs from your diet, you can still have cereal — use soy milk instead of dairy milk. I've been known to eat cold cereal with orange or apple juice on it. Other breakfast solutions for vegans are breads with fruit spreads or the tofu and bean spreads in this book.

Lunch

For lunch I usually enjoy soup and/or salad with bread, and some fresh fruit for dessert. Baked potatoes with any vegetable and yogurt topping are also great for lunch. If you're still eating meat, you can use your meat allotment at lunch in a sandwich. If you're a lacto-ovo vegetarian, you can have a slice of pizza or quiche, or an omelet. Vegans can try peanut butter and jelly or some variation on my favorite Dagwood: lettuce, tomato, onion, avocado, sprouts, and cucumber on whole wheat bread with honey mustard. Again, the spreads in this book, such as the Olive Creamy Cheese (page 242) or Hummos (page 247), are good for sandwiches. In a pinch you can grab a slice of pizza.

Of course, any of the soups, salads, side dishes, and entrées in this book can be used for lunch as well as dinner. Leftover anything from the night before makes a suitable lunch, too.

Dinner

Even the staunchest of meat eaters find certain meatless meals acceptable, such as spaghetti with marinara sauce, pasta primavera, vegetarian chili, cheese ravioli, eggplant Parmesan, curries, lo mein or noodles with sesame sauce, and guacamole and bean tostadas. You'll notice that most of these "mainstream" vegetarian dishes are of foreign origin. Perhaps the best way to ease into the habit of meatless meals is to start with foreign foods that are common main dishes, rather than trying to pass off a plate of steamed vegetables and brown rice as an entrée.

The recipes in this book do not contain any meat or fish, but you can adapt them to a low-meat diet if you are following the Modified plan. For example, cooked chopped meat or tuna, clams, or other seafoods can be stirred into any pasta sauce in this book, and you can certainly add chopped beef to any of the recipes for chili. Most of the Chinese dishes as well as the curries can accommodate the addition of meat or shellfish. By adding small amounts of meat to vegetarian dishes, you are still getting the benefit of eating less meat than you did before beginning the Modified plan. Two ounces of chicken may not seem like much when it's a lump on a plate, but it certainly goes far as part of a Chinese dish.

All the recipes in this book are suitable for lacto-ovo vegetarians. The vegan recipes — those without dairy products or eggs — are indicated by a (V) beneath the recipe title.

A MENU SAMPLER

Here are two weeks' worth of menus for anyone wanting to follow the New Four Food Groups program. All items given in CAPITALS AND SMALL CAPITALS are recipes included in this book. Almost all of these recipes are followed by one or both of the following symbols: an asterisk (*), which means it can be made in advance and frozen, and a (V), which indicates a vegan recipe.

❧ MENU 1 ❧

Prune Juice
PUMPERNICKEL (* V)
HONEY-ORANGE PEANUT BUTTER (* V)
Coffee or Tea

STUFFED BAKED POTATOES
STEWED CHICKPEAS AND OKRA (V)
Romaine Salad with Red Onion
Orange Wedges

JAMBALAYA RICE AND BEANS (V)
Steamed Cauliflower
NAKED TOMATO SALAD (V)
Raspberries and Sliced Peaches

❧ MENU 2 ❧

Orange Juice
Whole Wheat Bagel
LIPTAUER (*) or SESAME SPREAD (* V)

THREE-BEAN SOUP (* V)
Green Salad with Shredded Carrot, Sprouts, and Pumpkin Seeds
ANADAMA BREAD (* V)
Fresh Pineapple

FRAGRANT EGGPLANT WITH FRESH FIGS (V)
Brown Rice
Sliced Cucumbers
Fresh Pineapple

❧ MENU 3 ❧

Pineapple Juice
Toasted English Muffin with
HERB AND GARLIC CREAMY CHEESE
Coffee or Tea

OLD-FASHIONED SPLIT PEA SOUP (* V)
TOSSED SALAD WITH AVOCADO AND BLUE CHEESE DRESSING
MOLASSES-OATMEAL BREAD (*) or Whole-Grain Rolls
Golden Delicious Apple Slices

CREAM OF ASPARAGUS AND CARROT SOUP (*)
HERBED CHICKPEAS WITH AROMATIC RICE (V)
Chunky Celery, Tomato, and Olive Salad
GRAHAM BREAD (* V)
Fresh Bing Cherries

❧ MENU 4 ❧

Cantaloupe
CRAZY MIXED-UP CEREAL (V)
Coffee or Tea

GREEN GAZPACHO (V)
BEAN SPROUT AND WATERCRESS SALAD WITH GRAPEFRUIT-MINT
DRESSING (V)
CRANBERRY-PUMPKIN BREAD (*)
Bartlett Pear Wedges

MINESTRONE (* V)
VEGETABLE LASAGNE (*)
Romaine with Italian Dressing
Semolina Bread
Fresh Plums

❧ MENU 5 ❧

ORANGE-GRAHAM WAFFLES (*) or Cold Cereal with Soy Milk
Fresh Strawberries
Coffee or Tea

CORN CHOWDER (*) or VEGETABLE-BEAN SOUP (* V)
TOMATO, VIDALIA ONION, AND CHICKPEA SALAD (V)
BUTTERMILK BISCUITS (*)
Fresh Peach Slices

SENEGAL STEW WITH MILLET (V)
Steamed Wax Beans
Sautéed Greens
Mango Sorbet

❧ MENU 6 ❧

Cold Cereal with Soy Milk
Sliced Banana
Coffee or Tea

FRESH VEGETABLE SOUP (V)
TOSTADAS (V)
Seedless Red Grapes

SPAGHETTI SQUASH WITH VEGETABLE SAUCE (V)
Chilled Broccoli Vinaigrette
Sourdough Bread
Honeydew Melon

❧ MENU 7 ❧

Orange Juice
GRANOLA (* V)
Milk
Coffee or Tea

SWEET POTATO–RED LENTIL SOUP (* V)
WALNUT-RAISIN ROLLS (* V)
Endive and Watercress Salad
Raspberries or Blueberries

WHITE BEAN CHILI WITH WHEAT BERRIES (* V)
DICED VEGETABLE SALAD (V)
Corn on the Cob
Applesauce

❧ MENU 8 ❧

Cranberry Juice
PUMPKIN PIE WAFFLES (*) or Peanut Butter on Cinnamon Toast
Coffee or Tea

HUMMOS (* V)
Pita
Tossed Green Salad
TAHINI DRESSING (* V)
Fresh or Dried Apricots

SWISS CHARD SOUP (V)
POLENTA WITH EGGPLANT AND BLACK BEAN SAUCE (* V)
Steamed Green Beans
WILTED CUCUMBER SALAD (V)
THREE-GRAIN BREAD (* V)
Poached Pears

❧ MENU 9 ❧

Banana-Pineapple Juice
HONEY–WHOLE WHEAT PANCAKES or Toasted English Muffin
with TOFU FRUIT CHEESE (V)
Coffee or Tea

YUPPIE PASTA SALAD (V)
GARLIC BITES (* V)
Sliced Peaches

HOUSE-SPECIAL SOUP (V)
SPICY TOFU WITH CLOUD EARS (V)
FRIED RICE (V)
Orange Wedges

❧ MENU 10 ❧

SPICED FRUIT COMPOTE (* V)
EXTRA-HIGH-FIBER OATMEAL (V)
Coffee or Tea

WHEAT BERRY, ORANGE, AND BEAN SALAD (V)
Sliced Papaya

ALOO GOBI (V)
RED LENTILS AND CUCUMBER RICE (V)
Onion Relish
Fresh Pineapple

❧ MENU 11 ❧

Fresh Blueberries
Oatmeal or French Toast with Cinnamon Sugar
Coffee or Tea

ONION SOUP (* V)
Arugula and Endive Salad with GUILT-FREE ROQUEFORT DRESSING
PROVENÇALE SPREAD (V)
WHOLE WHEAT BAGUETTE (* V)
Ripe Nectarines

BAKED ZITI AND EGGPLANT WITH BASIL-TOMATO SAUCE
or
PENNE FROM HEAVEN (V)
Sautéed Sliced Zucchini
Romaine with Italian Dressing
Italian Bread
Ripe Pears

❧ MENU 12 ❧

Cottage Cheese with Strawberries and Orange Slices
GRANOLA (V)

SHAKER SUCCOTASH SOUP (*)
MARINATED CHICKPEA SALAD (V)
THREE-GRAIN BREAD (* V)
Granny Smith Apple Wedges

MEDITERRANEAN FAVA BEANS AND BULGUR (V)
Sautéed Spinach and Garlic
WHEAT BERRY BREAD (*) or Italian Bread
Honeydew and Cantaloupe Salad

❦ MENU 13 ❦

Pink Grapefruit
MOLASSES–RAISIN BRAN MUFFINS (*) or Cream of Wheat
Coffee or Tea

TROPICAL SALAD WITH CALYPSO DRESSING (V)
AMARANTH CORNBREAD (*) or Whole Wheat Rolls
Watermelon Wedge

DAL (* V)
CURRIED SPINACH, TOFU, PEAS, AND POTATOES (V)
RICE WITH EAST INDIAN FLAVORS (V)
Yogurt with Chopped Cucumbers
Chutney
Sliced Mango

❦ MENU 14 ❦

Fresh Blueberries
Lemon Yogurt or Toasted English Muffin with Peanut Butter
TEFF BANANA BREAD (*)

ASPARAGUS AND CHICKPEA SALAD (V)
RED PEPPER PIZZA BREAD (* V)
Kiwi Slices

ZITI WITH WHITE BEAN MARINARA SAUCE (V)
Steamed Asparagus
Fresh Spinach and Mushroom Salad
THREE-GRAIN BREAD (* V)
Fresh Apricots

2.

Cooking Basics

COOKING GRAINS AND BEANS —
BASIC KNOW-HOW AND CAVEATS

THE COOKING OF BOTH grains and beans is a matter of trial and error, no matter how exact or detailed the recipe may be. There are many variables that will cause basic instructions to yield different results from one time to the next. Some of these variables are discussed below.

Problem: Temperature. This is an obvious variable in cooking; if you cook at a higher or lower temperature than I do, naturally your cooking times will be different from mine. But how can we ensure that we're cooking at the same temperature? The setting for medium-high heat can vary from stove to stove, and it certainly varies even more when you are using a gas stove, which doesn't even have a specific setting for medium high.

Solution: Periodically lift the lid to make sure that the water is simmering or cooking at the described pace. For grains, start checking for doneness 10 to 15 minutes before the recipe states they will be cooked, and for beans, start checking 30 minutes ahead.

Problem: Type of cookware used. Heavy cookware conducts heat more evenly than thin cookware, and although you may think that would speed up the rate of cooking, in fact it slows it, so that overcooking doesn't occur in one area of the pan while other spots are undercooked. If your cookware is thin, you can use a simmering pad — a round metal plate made of two layers of perforated metal — that will disperse the heat evenly. Simmering pads can be found in houseware departments and sometimes in hardware stores.

The fit of the lid is also crucial. Very tight-fitting lids may keep all the moisture in the pot, in which case you may need a little less water than a recipe calls for. The need to use less water is especially common for grains, which are supposed to cook until tender, at which point all the liquid should be absorbed. The amount of cooking liquid used with beans is less critical, since you expect it to be drained off when the beans are finished and, more important, since you usually cook beans loosely covered. Conversely, loose-fitting lids will allow more steam to escape and may cause the water to evaporate before the grains (or beans) are tender.

Solution: Know your cookware. Try to use the same pot or type of pot for all your basic grain and bean cooking, so you can become familiar with its peculiarities. If there is always extra water left after grains are cooked, plan to start out with less than the basic recipe calls for in the future, and for now just drain off any extra cooking liquid. If, on the other hand, there is always too little water to cook the grains until tender, you may want to plan to begin with a little extra. To remedy this problem while you are cooking the grains, add a little extra boiling water to the pot and let the grains continue to cook.

Problem: Age of grains and beans. This variable significantly affects the cooking time for beans and, to a lesser extent, grains. If beans are this year's harvest, they will require less cooking time than older beans, and frequently will not even have to be soaked. (The cooking method for new beans is to cover with water and bring to a boil, cook 2 minutes, drain and cover with fresh water; then bring to a boil again and cook until tender.) The problem is that beans do not come dated (with the exception of a few gourmet beans, such as those sold by Dean and DeLuca).

Solution: One clue to the age of beans is their color and luster. Are the beans faded and dull and/or shriveled? Then they are probably old. They are still acceptable for cooking, since the nutritional value is unchanged, although they will tend to fall apart more easily during cooking and their cooked texture may be less pleasing than that of fresher beans. Bright colors and some shine will indicate newer beans. When in doubt, your best bet is to soak your beans before cooking (see explanation for soaking, pages 48–49). Start to check for doneness halfway through the specified cooking time.

Grains are affected to a lesser degree; start checking for doneness 10 to 15 minutes before the end of the prescribed cooking time.

Problem: Hard water. Because hard water has more mineral salts ("salt" is the operative word) than normal water, beans may not cook properly in hard water. It may slow the cooking time or even prevent cooking completely.

Solution: A pinch of baking soda — absolutely no more than ⅛ teaspoon per cup of beans cooked, or the beans will be too mushy — added to the water will counteract the mineral salts.

Problem: High-altitude cooking. Beans and grains will take longer to cook at high altitude.

Solution: Check with your local agricultural extension offices for information on high-altitude cooking conversion. All the recipes in this book will need to be converted, since they were developed at sea level.

COMMONLY ASKED QUESTIONS ABOUT GRAINS AND BEANS

The following are, in my experience, the questions most frequently asked by people who are trying to include home-cooked grains and beans in their diet. A careful review of this section should enable both the beginning and experienced cook to approach the recipes with confidence.

How Can I Prevent Gasiness from Beans?

The culprit for that uncomfortable feeling after eating beans is sugars called oligosaccharides. Your intestines are not well equipped to digest these sugars, so there they sit — fermenting trouble.

One solution is to buy a product called Beano Drops, which you put on your first bite of offending food (not just beans but also cabbage, broccoli, brussels sprouts, or any other vegetable that gives you gas). It works quite well; look for it in health food stores, Kmart, Walgreen's and Wal-Mart, or mail order it from The Bean Bag (see "Sources," page 63).

Another way to beat gas is to eat beans frequently. Most people who eat them regularly don't have a problem with gas. (I find this true, with the exception of undercooked beans, which still do me in!) You may want to accustom your body to beans by starting out with small portions, building your stamina gradually.

I find that canned beans are less gas-producing than home-cooked. One reason is that they're never undercooked. A second reason has to do with the fact that they are packed in salted liquid. Salt does improve the digestibility of beans, but it also interferes negatively with proper cooking — so homemade beans, which cook better when salt is added *after* cooking, don't absorb enough salt during this short period to improve their digestibility. Canned beans, on the other hand, absorb the salt from the canning liquid. Draining and rinsing canned beans before using them in recipes will get rid of some of the salt but will leave the antigas benefits.

There are some other ways to address the gas problem. Changing the soaking water two or three times is one. The sugars responsible for gas are water-soluble, so be sure to throw out the soaking water and rinse the beans, then repeat once or twice more during the soaking period, before starting the cooking process with fresh water. Similarly, discard the cooking water after the beans are tender (unless, of course, you are cooking the beans in a soup or stew). Another method is to cook the beans thoroughly. The less cooked the bean, the harder it will be to digest.

Are Beans Poisonous?

Two of the many sources that I've used in researching this book state that beans contain toxins called lectins, which can cause diarrhea, stomach cramps, and nausea. In order to detoxify the lectins, these sources state, you must boil your beans for 10 minutes (the high heat will destroy them). None of the other books on beans I've consulted mention this, and in fact, most cookbooks suggest that beans be simmered, not boiled, to achieve the best cooked consistency. Since simmering occurs at almost the same high temperature as boiling, it is also effective in eliminating lectins.

I skip the boiling and use the simmering method to cook beans. I have yet to meet anyone who has developed the symptoms described above from eating simmered beans. But this is entirely up to you. If you feel safer boiling your beans before lowering the heat and simmering them, I'm sure the quality of the final product will be fine.

Should I Soak My Grains and Beans?

Some cookbooks do advise soaking grains before cooking them. I honestly don't know why. Grains cook just fine without any soaking. The one exception is glutinous rice, which should be soaked before cooking.

For most beans, soaking is usually a good idea (the exceptions are lentils, split peas, and sometimes black-eyed peas). One reason is that the sugars that tend to give people gas are water-soluble and therefore can be minimized by soaking the beans and then discarding the soaking water. The second reason is that beans do not absorb water through their skins, but only through the hilum, where the bean was attached to the pod. By soaking the beans you allow enough time for the water to be completely and evenly absorbed.

To determine whether beans are adequately soaked, cut one in half. If the core of the bean has a lighter color than the rest of it, it is undersoaked; a properly soaked bean displays a uniform color.

Should I Use Salt in the Cooking Water?

For grains: Some whole grains — amaranth, pot or hull-less (not pearl) barley, whole-grain triticale, wheat berries (whole-grain wheat), and wild rice — do not absorb liquid properly when the water is salted. These grains must be salted *after* cooking. You may want to consider adding the salt to all your grains after cooking rather than before, because doing so will make them taste saltier. In this way you can use less salt to achieve a desired saltiness, which is important for anyone watching sodium intake.

For beans: Salt or anything acidic in the water prevents the beans from softening properly. Therefore, beans should be soaked and cooked in unsalted water — and this means they should not be cooked in broths, which are usually salty, or with tomatoes, lemon juice, or other acidic fruits or vegetables. Salt and other ingredients that would toughen the beans early in the cooking can be added after the beans have reached the desired tenderness.

How Can I Tell When My Grains/Beans Are Cooked?

In both instances the answer is: when they have the chewiness that you like. Doneness is definitely a matter of taste.

There are, however, a few general guidelines to help you decide doneness.

For grains: Ideally, grains are cooked when all the cooking liquid has been absorbed. But due to the variables previously mentioned (temperature of the stove, type of cookware, and age of grains), this test for doneness is not too reliable. To determine doneness, taste a few grains — they should have an al dente consistency. If they crunch at all, they are undercooked. Let them continue cooking and, if necessary, add extra boiling water. If the grains are mushy or soggy, they are overcooked.

For beans: If the beans are cooked, you should be able to mash one by pressing it against the roof of your mouth with your tongue. Another indication of doneness is that a few beans will have split their skins. Too many split skins can indicate that you have overcooked the beans or had the heat too high. If none of the beans have split their skins, they are probably still slightly undercooked. Biting into one bean will be the best test of all.

What's the Difference Between Canned Beans and Beans Cooked from Scratch?

I want to state right up front that I find canned beans to be excellent products. Although there is a difference in quality from one brand to the next, all but the very worst brands are generally quite acceptable, or even excellent, for use in these recipes.

There are, however, advantages to both canned and home-cooked beans. When you cook beans from scratch you have control over how much salt you want to use (canned beans are almost always packed in salted water). You also have a greater variety of beans from which to choose. For the most part, the kinds that are found canned are kidney beans, small white beans, cannellini, chickpeas, black beans, black-eyed peas, pink beans, Roman beans, butter beans, lima beans, and fava beans. In contrast, your choice of beans from scratch includes all these plus dozens more — many of which are absolutely delicious (see the mail-order sources, pages 63–64, for places where you can obtain some of the more exotic varieties).

Canned beans, on the other hand, are incredibly convenient. While you have to plan hours in advance if you want to make beans from scratch, canned beans are ready whenever you are — assuming you have them in your pantry.

Nutritionally, the major difference between canned and fresh-cooked beans is the salt content. This factor in turn affects digestibility, as discussed earlier.

Most of the recipes in this book are prepared with cooked beans that are also available canned. This way you have the choice between cooking from scratch and opening a can.

When a recipe calls for . . .	*you can use a . . .*
1 cup cooked beans	10½-ounce can, drained and rinsed
1½ cups cooked beans	16-ounce can, drained and rinsed
1¾ cups cooked beans	20-ounce can, drained and rinsed

What Qualities Make Some Canned Beans Better Than Others?

I look for the following things in canned beans: the flavor should be fresh, not tinny; the texture should be soft but not mushy; most of the beans should be intact, with some but not too many skins split; the canning liquid should be liquid, not gooey; and there should be little or no sludge at the bottom of the can.

The national brands that I find best are S & W and Green Giant. I have found some smaller, imported brands to be quite good too, so I suggest that you try a few brands until you discover your favorites.

Are Grains Interchangeable?

Grains that have similar textures and cooking times are frequently interchangeable.

The following whole grains can be used interchangeably when recipes call for them cooked: whole-grain wheat, rye, triticale, and oats.

Although different white rices have different flavors, all long-grain white rice is interchangeable. You can substitute cooked brown rice for cooked white rice as well. Cooked wild rice can also be used in place of some or all other cooked rice.

Cooked couscous can be substituted for cooked millet, but not the other way around, since the flavor of millet is so much stronger than that of couscous.

Bulgur and cracked wheat, which are both made from whole-grain wheat, are completely interchangeable. Cracked wheat, as the name indicates, is simply grain cracked into coarse pieces, with the bran and germ left on the grain. In the manufacture of bulgur, the bran is removed from the grain, and the grain is then steamed, dried, and cracked. Cracked wheat is composed of dark and light pieces, whereas bulgur is a uniform brown. The cooking times and yields for cracked wheat and bulgur are similar, as is the flavor. Cooked couscous can be substituted for either of them, and vice versa.

As for flours, you can substitute all-purpose flour for bread flour, but not the other way around. Although you can substitute all-purpose flour for whole wheat, rye, or oat flour, you'll lose the nutritional and flavorful qualities of the alternate flours. Whole wheat flour and graham flour are interchangeable. If you want a more authentic substitute for graham flour, replace 1 tablespoon of whole wheat flour per cup with 1 tablespoon of wheat germ. Any other substitutions among flours are not recommended because the changes in texture or flavor would be significant.

Are Beans Interchangeable?

Yes and no. My basic rule is, if two beans are of a similar size and shape, with similar cooking times, they can be used interchangeably. Some beans, however, do have distinct properties and are best used only when specifically called for.

The first category — interchangeable beans — is composed of what I call bean-shaped beans: those with great similarities in shape, size, and taste. Examples are Great Northern, adzuki, red, and pink beans, and black-eyed peas. Although some are more earthy-flavored and others slightly ashy, some vaguely sweet and others smoky, if you popped one of these beans in your mouth and had to say what you were tasting, your answer would probably be simply "Beans." Unless you were an expert — or were able to peek at the beans — you'd be unlikely to distinguish a kidney bean from a cannellini, pinto bean, or cranberry bean. The same is probably true of a black bean and a white bean.

The second category — noninterchangeable beans — is made up of those beans with distinctive tastes and cooking qualities that set them apart from beans in general. They include:

Lentils — have a peppery flavor, require no soaking, and have a short cooking time, relatively speaking. You can use green, brown, and French lentils interchangeably, although the green and French lentils cook slightly more quickly than the brown (the ones usually sold in the supermarket). Red lentils are entirely different and should not be substituted for the other lentils, or vice versa. I don't recommend substituting lentils for other beans.

Chickpeas — have a unique flavor and shape. They tend to require longer cooking than most other beans. Use your discretion in substituting them for other beans.

Lima beans — are one of the few beans that evoke strong opinions; many people either love them or hate them. Canned or frozen limas are not suitable substitutes for dry lima beans (although you can substitute frozen lima beans for cooked *flageolets,* which are delicate, and costly, dried beans). Canned butter beans are a better substitute for dried limas.

Fava beans — have a lovely flavor and texture. However, the outer skin of dried favas is so tough — even when properly cooked — that they must be peeled after cooking. This factor makes it difficult to substitute them directly for other beans. Fresh fava beans may or may not need peeling after cooking, depending on their age.

Split peas — need no soaking before cooking and tend to dissolve when cooked, because they are skinless. For this reason, the only substitutes that are suitable are green split peas for yellow split peas and vice versa. Red lentils, which are also skinless, have the same characteristic of dissolving and can be used in recipes calling for split peas, but they do cook much faster and have a different flavor.

What Is the Stovetop Cooking Method for Beans and Grains?

If you don't have a microwave oven or a pressure cooker, or if you simply feel more comfortable doing your cooking on the stove, the following charts provide complete information on the stovetop method. I've tried to include all the grains and beans you're likely to encounter.

⬿ INSTRUCTIONS FOR STOVETOP COOKING OF BEANS ⬿

1. Rinse the beans and discard any debris.
2. For each cup of beans, add 4 cups of water and soak overnight, or until the interior of the bean is uniform in color when cut in half with a sharp knife. Or quick-soak by bringing the water and beans to a boil, boiling 2 minutes, and letting stand 1 hour, or until the interior of the bean is uniform in color when cut in half with a sharp knife.
3. Discard the soaking water.
4. Place 4 cups of fresh water and the soaked beans in a 2-quart saucepan and bring to a boil. Reduce the heat and simmer, covered loosely, for the suggested cooking time given below.
5. Start checking for doneness at the lower cooking time.

Type of Bean	Cooking Time
Adzuki	¾ to 1½ hours
Black	1 to 1½ hours
Black-eyed peas	¾ to 1½ hours
Cannellini	1 to 1½ hours
Chickpeas	2 to 3 hours
Fava (peel after cooking)	2 to 3 hours
Great Northern	1 to 2 hours
Kidney	1 to 2 hours
Lentils	½ to 1 hour
Lima (baby)	¾ to 1½ hours
Lima (large)	1 to 1½ hours
Mung	¾ to 1 hour
Navy (small white)	1 to 2 hours
Pink	1 to 2 hours
Pinto	1 to 2 hours
Roman (cranberry)	1 to 2 hours
Soy	2½ to 3½ hours

❧ INSTRUCTIONS FOR STOVETOP COOKING OF GRAINS ❧
FOR 1 CUP UNCOOKED GRAIN.

1. Rinse the grain and discard any debris.
2. Bring the water to a boil in a 2-quart saucepan.
3. Stir in the grain and, unless otherwise noted under Comments, the salt (if desired).
4. Reduce the heat and simmer, covered, for the suggested cooking time given below.
5. Let stand for the specified time.
6. Test for doneness.

Grain	Add Water	Stir in Salt	Cook Covered*	Let Stand	Yield	Comments
Amaranth	1½ cups	½ tsp.	20 min.	†	2 cups	Add salt *after* cooking
Barley, hull-less	2½ cups	1 tsp.	1⅓ hours	10 min.	4¼ cups	Add salt *after* cooking
Barley, pearl	2½ cups	1 tsp.	40 min.	5 min.	3½ cups	
Barley, pot	2½ cups	1 tsp.	1⅓ hours	10 min.	3¼ cups	Add salt *after* cooking
Bulgur or cracked wheat	2 cups	¾ tsp.	20 min.	5 min.	3 cups	
Couscous	1¾ cups	½ tsp.	1 min.	5 min.	2¾ cups	
Millet	2 cups	½ tsp.	25 min.	5 min.	4 cups	Sauté in oil until lightly browned before adding to water
Kasha (roasted buckwheat)	2 cups	1 tsp.	15 min.	5 min.	3½ cups	Toss in beaten egg and cook until dry before adding water
Oats, whole	2 cups	½ tsp.	1 hour	10 min.	2½ cups	
Quinoa	2 cups	1 tsp.	15 min.	5 min.	3½ cups	Rinse very thoroughly before cooking
Rice, brown	2¼ cups	1 tsp.	50 min.	5 min.	3½ cups	
Rice, white	2 cups	1 tsp.	20 min.	5 min.	3 cups	Don't rinse if enriched
Rye, whole	2½ cups	½ tsp.	2 hours	10 min.	3¼ cups	
Teff	3 cups	½ tsp.	15 min.	†	3 cups	Don't rinse; stir occasionally
Wheat, whole	2¼ cups	½ tsp.	1¾ hours	10 min.	2½ cups	Add salt *after* cooking
Wild rice	2 cups	¾ tsp.	55 min.	5 min.	3 cups	Add salt *after* cooking

Cooking times may vary depending on cookware used, cooking temperature, and age of grain. Check for doneness at least 10 minutes before prescribed cooking time.

†*Do not let stand; the grain will become stiff.*

Can I Microwave Grains and Beans?

Grains cook very well in the microwave. But for the grains that require shorter cooking times (less than 30 minutes), such as white rice, bulgur, couscous, and quinoa, the time saved is minimal. The longer-cooking grains, such as whole-grain wheat, rye, or oats, cook much faster in the microwave. In all instances the amount of water used to cook the grain differs slightly from that called for in stovetop recipes. Consult the table "Instructions for Microwaving Grains," below, for complete information.

❧ INSTRUCTIONS FOR MICROWAVING GRAINS ❧
FOR 1 CUP RINSED, UNCOOKED GRAIN.

1. Place the specified amount of water in a large microwave-safe bowl, cover with waxed paper, and microwave on high for the specified time.

2. Stir in the grain, re-cover, and microwave on medium for the specified time, rotating the bowl once or twice if necessary.

3. Let stand for the specified time.

4 Stir in the salt, if desired.

Grain	Measure Water	Microwave, Covered, on High Power	Microwave, Covered, on Medium Power	Let Stand	Stir in Salt	Yield
Barley, pearl	2½ cups	5 min.	35 min.	5 min.	¾ tsp.	3½ cups
Bulgur or cracked wheat	2 cups	3 min.	15 min.	2 min.	¾ tsp.	3½ cups
Couscous	1½ cups	4 min.	1 min.	3 min.	½ tsp.	2¾ cups
Kasha (roasted buckwheat)	1¾ cups	3 min.	15 min.	5 min.	¾ tsp.	3¼ cups
Millet*	1¾ cups	5 min.	20 min.	4 min.	½ tsp.	3½ cups
Oat groats	3 cups	7 min.	1 hour	5 min.	½ tsp.	2¾ cups
Quinoa†	2 cups	3 min.	15 min.	5 min.	¾ tsp.	3 cups
Rice, brown	2½ cups	5 min.	42 min.	3 min.	¾ tsp.	3 cups
Rice, white	2¼ cups	4 min.	20 min.	2 min.	¾ tsp.	3 cups
Rye berries	3 cups	7 min.	1 hour	5 min.	1 tsp.	2¼ cups
Wheat berries	3 cups	7 min.	1 hour	5 min.	½ tsp.	2¼ cups
Wild rice	2½ cups	5 min.	50 min.	5 min.	¾ tsp.	3¼ cups

*Stir the millet and 1 tablespoon oil together, microwave uncovered on high 2 minutes, stir, and microwave uncovered on high 2 minutes longer. Then stir in 1¾ cups hot water and proceed according to directions above.

†Rinse very thoroughly, until soapy bubbles stop forming.

Technically, it is possible to microwave beans. After sorting and rinsing the beans, place them in a large microwave-safe bowl, add 4 cups water for each cup of dried beans, cover, and microwave on high (100 percent) power until boiling. Simmer on medium (50 percent) power for 2 minutes and then let stand 1 hour. Drain, rinse, and add 4 cups more water for each cup of dried beans you started out with. Cover and microwave on high power until boiling. Cook at medium power until tender.

Should you microwave beans? I have found that it takes just about the same amount of time to microwave beans as it does to cook them on the stovetop, and they don't cook as evenly in the microwave. So why bother?

Can I Use a Pressure Cooker for Grains and Beans?

I should preface this section with the statement that everything I know about pressure-cooking I learned from Lorna Sass, author of *Cooking Under Pressure* (William Morrow, 1989), who was kind enough to hold my hand and show me how to work that frightening piece of equipment. Before Lorna educated me, I had owned two pressure cookers for more than three years and never dared to try either one, because all I could think of were the numerous times my mother's pressure cooker had exploded and splattered food on our kitchen ceiling.

Pressure cookers do, in fact, significantly reduce the amount of time needed for cooking grains and beans. But this blessing is not without its drawbacks. As you've already read, each batch of grains and beans (especially beans) can vary in the amount of cooking time necessary, due to differences in age of grain or bean, cooking equipment used, and temperature of the stovetop. These factors also affect times for pressure-cooking. The problem, then, is that unless you actually stop the cooking process completely, there is no way to check for doneness. If you do stop the pressure cooker and find that the grains or beans are slightly undercooked, by the time you return it to the right pounds of pressure needed to continue cooking, you may have already over-cooked the item in the cooker. The best method for using the pressure cooker is to cook the grains or beans only to the minimum recom-mended cooking time. If the grains or beans are undercooked, complete the cooking on the stovetop without pressure until desired doneness is achieved.

The charts on pages 57 and 58 are abridged from *Cooking Under Pressure*.

Instructions for Pressure-Cooking Grains and Beans

1. Sort and rinse the grains or beans.

2. Soak overnight in 4 cups water for each cup of dried beans (skip this step for grains).

3. Discard the soaking water and any loose or free-floating skins (these can clog the vent and create excessive pressure buildup).

4. Use 4 cups water and 1 tablespoon vegetable oil for each cup of dried grains or beans. (The addition of the oil helps prevent clogging of the vent — so don't omit it because you want to reduce your fat intake!)

5. Do not fill the pressure cooker above the halfway mark.

6. Lock the lid, bring to high (15 pounds) pressure (consult your user's manual for instructions), reduce the heat (usually to medium-low or whatever heat is necessary to maintain pressure), and cook for the suggested length of time.

7. Quick-release pressure by placing the cooker under cold running water — tilt the lid away from you when you open the pot, to avoid being burned by the steam.

8. Check for doneness. If the beans are done, drain. If they are not quite cooked, complete the cooking without pressure on the stove-top. If the beans are significantly undercooked, return the pot to high pressure and continue cooking.

9. Add salt, if desired.

❧ PRESSURE-COOKING TIMES FOR GRAINS AND ❧ BEANS — MINUTES UNDER HIGH PRESSURE

Item	*Minutes*
Black beans	9 to 11
Cannellini	9 to 12
Chickpeas	10 to 12
Lentils (don't need to be soaked overnight)	7 to 10
Navy beans	6 to 8
Pinto beans	4 to 6
Red kidney beans	10 to 12

Item	Minutes
Barley, pearl	18 to 20
Oat groats	25 to 30
Rye berries	25 to 30
Wheat berries	35 to 45
Wild rice	22 to 24

Some grains are cooked by using "natural pressure release." This means cooking briefly under high pressure and then letting the grain stand in the pressurized pot, without heat, for a specified number of minutes. The pot is then placed under cold running water until it stops steaming and is cool enough to handle.

◄ INSTRUCTIONS FOR NATURAL PRESSURE ◄ RELEASE COOKING

FOR 1 CUP RINSED, UNCOOKED GRAIN. FOLLOW THE DIRECTIONS IN THE SECOND, THIRD, AND FOURTH COLUMNS, MOVING FROM LEFT TO RIGHT.

Grain	Add Water	Cook Under High Pressure	Let Stand in Pot
Quinoa	2¼ cups	2 min.	10 min.
Rice, brown	1¾ cups*	15 min.	10 min.
Rice, white	1½ cups	3 min.	7 min.

Start with boiling water.

◄

BROTH: THE FLAVOR ENHANCER

Broth lends a desirable complexity to the flavor of cooked foods, a complexity you will miss if you choose to substitute water for broth (one option) in the recipes in this book. In fact, people not used to vegetarian cooking often find some dishes too bland, largely because they are made without chicken or beef broth. Many of the recipes call for vegetable broth, but give the option of chicken or beef. If you have not become a vegetarian, I would suggest that you opt for canned chicken or beef broths — they have a high-quality flavor and are extremely convenient.

Unfortunately, there is no canned vegetable broth currently on the market. For the vegetarian, there are two choices: cooking your own broth from scratch or using some form of dry or semidry vegetable bouillon. I've tried to taste as many of these vegetable broths as possible, and I have found — to be perfectly honest — that they range in flavor from fairly good to absolutely vile. Here are descriptions of the five that I liked best, in order of preference. If a vegetable broth is not included below, it's either because I tried it and didn't like it at all, or because it wasn't available in my area. Of course, since this is a matter of taste, you should feel free to try other brands; you may find you like them better than I did.

Knorr's Vegetarian Vegetable Bouillon. These vegetable cubes come in a small yellow box. The cubes are large, and each one makes 2 cups of broth. The flavor is mellow and rounded, and only slightly salty. The broth is golden and is a suitable replacement for chicken broth. It does contain palm oil, which is high in saturated fat.

Romanoff MBT Instant Vegetable Broth. This broth comes in foil packets in a green cardboard box. It's very beefy in flavor and very salty. You may want to dilute it more than recommended. The same firm also puts out a vegetarian onion broth that has similar taste qualities, except that it also tastes oniony.

Carmel Parve Instant Soup Mix. The label on this jar of bouillon powder states "tastes like beef soup." Not quite — its flavor falls somewhere between beef and chicken, with a heavy accent of onion. The broth is clear and the taste is a little salty but not bad. It has the added advantage of being certified kosher, in case you follow Jewish dietary laws. Look for it in the kosher department of your supermarket rather than with the soups.

Veget-All Soup. This yellow powder, which comes in a glass jar, tastes remarkably similar to dry powdered chicken broth, the kind found in packets. The broth is very yellow, cloudy, and slightly powdery, with a hint of soapiness. It's available in health food stores, and its biggest asset is that it comes in a salt-free version.

Barth's Nutra Soup is a powder that comes in a glass jar. It's available in various flavors. I found the vegetable flavor better than the celery, carrot, or onion. The broth is cloudy and greenish, with pieces of herbs and other stuff floating in it. It has a grassy flavor and slightly powdery mouth feel. Its big plus is that it's unsalted. Available through natural food stores.

Definitely the best vegetable broth is the one you make yourself. Unfortunately, to get a really good broth you have to use lots of vegetables, and that can run into money. The good news is that since broth freezes well, you can make it in large batches and then freeze them in 1-cup portions. You can store the broth in plastic containers or use either a heat- or vacuum-sealing system.

Here's my favorite vegetable broth recipe. I developed it for my first book, *The Complete Whole Grain Cookbook* (Donald I. Fine, 1989), and I've never come up with a better one.

☙

VEGETABLE BROTH

12 cups (3 quarts) water
¼ head cabbage (about ½ pound)
4 medium carrots, peeled
4 ribs celery
l 3 medium leeks (white and light green parts only), thoroughly rinsed
3 medium parsnips, peeled
2 medium kohlrabi, peeled
2 white turnips, peeled
1 large tomato
1 large onion, peeled
1 small celeriac (celery root), peeled
1 bunch Italian parsley (flat leaf)
1 tablespoon fresh lemon juice
¾ teaspoon salt

Place the water, cabbage, carrots, celery, leeks, parsnips, kohlrabi, turnips, tomato, onion, and celeriac in a 6- or 8-quart stockpot. Bring to a boil over high heat. Reduce the heat and simmer, uncovered, 1¾ hours. Add the parsley, lemon juice, and salt. Simmer 30 minutes longer. Strain the broth through a colander, pressing the vegetables against the colander with the back of a spoon to extract all the flavors before discarding.

MAKES 5 CUPS

☙

STOCKING YOUR PANTRY

SOMETIMES SUCCESS in following the New Four program will depend on your being able to put together a last-minute meal quickly.

That in itself is not a problem — as long as you have the ingredients you will need. The following is a list of the products that you should keep in your pantry, refrigerator, and freezer. This list represents a minimum; I usually keep much more on hand.

BEANS:

Canned:	*Dried:* Any varieties you like.
black beans	Some basics are:
cannellini	black beans
chickpeas	black-eyed peas
kidney beans	kidney beans
white beans	lentils
Prepared: peanut butter	lima beans
Frozen: lima beans	small white beans
	split peas

GRAINS:

barley	oat bran
bulgur or cracked wheat	oatmeal
cornmeal	pasta, all sizes and shapes
flour, all-purpose and whole wheat	rice, brown and white

VEGETABLES for Use in Cooking:

bell pepper, green and/or red	parsley and any other fresh herbs you like (basil, coriander, dill)
carrots	
celery	
garlic	parsnips (good for soup making)
leeks (good for soup making)	
onions	potatoes
	scallions
	tomatoes

FRUITS:

Fresh:	*Dried:*
apples	raisins
lemons	
everything that you love	

NUTS AND SEEDS:

almonds   poppy seeds
caraway seeds   sesame seeds
pecans   walnuts
pignoli (pine nuts)

SPICES:

chili powder   ginger, ground and fresh
cinnamon   paprika
curry powder   red pepper, ground

DRIED HERBS:

basil   oregano
bay leaf   thyme
dill

CANNED GOODS:

broth, chicken or beef   tomato sauce
corn kernels   whole peeled tomatoes
tomato paste

STAPLES:

baking powder   olives, green or black
baking soda   Red Hot sauce or Tabasco
bouillon cubes/powder, vege-   salt
   table or other   soy sauce
cornstarch   sugar
honey   tahini
ketchup   vinegar, red wine and distilled
mirin or dry sherry   white
mustard   Worcestershire sauce
oil — olive, sesame, and vege-
   table

FROZEN FOODS:

peas
spinach, chopped

༨

SOURCES

HERE ARE SOME SOURCES of mail-order ingredients you may not be able to find locally.

Arrowhead Mills
P.O. Box 2059
Hereford, TX 79045

Grains, cereals, beans

Balducci's
11–02 Queens Plaza South
Long Island City, NY 11101

Gourmet prepared foods, fresh pasta, meats, cheeses, fish

The Bean Bag, Inc.
818 Jefferson Street
Oakland, CA 94607

The largest array of unusual dried beans available, Beano Drops

Dean and DeLuca
560 Broadway
New York, NY 10012

Dried beans, grains, unusual produce, gourmet foods, cooking equipment, cookbooks

Eden Foods, Inc.
701 Tecumseh Road
Clinton, MI 49236

Grains, beans, baked goods, Japanese products, pastas, snacks, lima products from Belgium

Frieda's Finest/Products Specialties, Inc.
P. O. Box 58488
Los Angeles, CA 90058

"New and Unusual Produce" — baby and other vegetables, fruits, herbs, lettuce, marinated items, fresh and dried mushrooms

Walnut Acres
Penns Creek, Pa 17862

Grains, beans, prepared foods, baked goods, equipment (including grain mills)

Many of the ingredients mentioned in this book can be found in natural or health food stores. You can also find many unusual ingredients in specialty ethnic stores, such as Oriental, Indian, and Middle Eastern shops.

3.

The Recipes

BEFORE YOU DIVE INTO the recipe chapters — Soups, Main Dishes, Side Dishes, Salads, Breads and Spreads, and Breakfast Dishes — here are a few things you should know.

Most of the recipe titles are followed by one or more of these three symbols: (E), (Q), and (V).

(E) indicates an easy recipe, one requiring little work or cooking expertise to prepare. Some (E) recipes call for extended cooking time (such as soups, in which assembly will be simple but simmering time long), so that all (E)s are not necessarily speedy.

(Q) denotes a recipe that takes less than 30 minutes to prepare, from start to finish. Because in some cases I have assumed that you are going to cook the beans from scratch, there will be no (Q) following the recipe title. If, however, you choose the canned bean option, that recipe may become (Q) — and (E) as well.

(V) indicates a vegan recipe and hence contains no meat, eggs, or dairy products. When a recipe with a (V) includes margarine, I'm assuming that vegans will choose a brand made exclusively from vegetable oils. Some (V) recipes may have an optional cheese ingredient, such as Parmesan, but all are extremely good without the cheese. Similarly, recipes with optional ingredients like chicken broth or milk, given in parentheses after a vegan ingredient, are also considered (V). I have also labeled (V) a few Indian recipes that call for ghee (clarified butter), because vegetable oil can easily be substituted for it.

At the end of the recipe instructions you will find the yield of the recipe. Serving sizes usually mean 1 cup per person for soups and entrées, and about ½ cup per person for side dishes and salads (except leafy salads, which provide 1-cup servings). If a recipe gives a range of servings, such as "Serves 4 to 6," this means that it makes 4 generous servings or 6 smallish servings. Unless otherwise indicated, 4 servings means one quarter of the total yield for the recipe, and 6 servings one sixth. Servings for breads are 1 slice. As I note earlier, most entrées can be served as side dishes and vice versa, simply by altering portion sizes. Salads, many of which contain satisfying ingredients, can also be converted into main dishes, by increasing the serving size.

The last item in each recipe is the nutritional analysis, which gives approximate figures for calories, protein, carbohydrates, fiber, total fat, saturated fat, cholesterol, and sodium. These figures should be considered guidelines rather than absolutes. All calculations are based on the highest-fat form of ingredients and the largest-size portion: for example, if a recipe calls for milk, whole milk was used in making the calculations, and if a recipe serves 4 to 6, the nutritional information is based on 4 servings. These figures also include all optional ingredients. Obviously, if you use a lower-fat product than the one called for in the recipe, and/or a smaller serving size, your version will contain fewer calories and less fat, saturated fat, and cholesterol than indicated in the nutritional analysis (see the chart on page 19 for a detailed breakdown).

A reminder: I have created the recipes to match my own taste for spicy foods. If you like milder flavors, use half or less of the pepper or pepper sauce that I call for. My brand of hot sauce is Durkee Red Hot, which is not as hot as Tabasco. If you are using Tabasco, you may want to use smaller amounts than those called for.

Throughout the recipes, I have called specifically for precise cup measures of chopped vegetables and nuts, and of cooked beans, which you can either make at home or buy canned. For your convenience, the following two lists give you an idea of how much to buy to equal the amounts required in the recipes.

RECIPE EQUIVALENTS FOR VEGETABLES AND NUTS

Amount Required in Recipe (Trimmed Weight)	Equals	Amount Whole Vegetable or Nut
1 cup asparagus cut into 1-inch pieces (3½ ounces)	=	5 medium asparagus
2 cups beans (any kind), dried (16 ounces)	=	1 pound dried beans
3½ cups broccoli florets (9 ounces)	=	1 medium bunch broccoli
4½ cups shredded cabbage (9 ounces)	=	½ small head cabbage
3 cups chopped cabbage (9 ounces)	=	½ small head cabbage

Amount Required in Recipe (Trimmed Weight)	Equals	Amount Whole Vegetable or Nut
1 cup sliced carrot (6 ounces)	=	1 large carrot
1 cup coarsely chopped carrot (6 ounces)	=	1 large carrot
1 cup finely chopped carrot (6 ounces)	=	1 large carrot
1 cup shredded carrot (6 ounces)	=	1 large carrot
5 cups cauliflower florets (20 ounces)	=	1 medium head cauliflower
1 cup sliced celery (4 ounces)	=	2 medium ribs celery
1 cup chopped or diced celery (4½ ounces)	=	2 medium ribs celery
1 cup finely chopped or diced celery (4¾ ounces)	=	2 medium ribs celery
1 cup chopped, sliced, or diced cucumber (5½ ounces)	=	½ large cucumber
5 cups cubed eggplant (14 ounces)	=	1 medium eggplant
1 cup green beans cut into 1-inch pieces (4 ounces)	=	⅓ pound green beans
3 tablespoons sliced or chopped scallion, white and green parts (½ ounce)	=	1 medium scallion
2 tablespoons finely chopped scallion, white and green parts (½ ounce)	=	1 medium scallion
1 cup sliced leek, white and light green parts only (2½ ounces)	=	1 medium leek
1 cup sliced or chopped mushrooms (3 ounces)	=	4 medium mushrooms
1 cup chopped nuts	=	4 ounces shelled nuts
1 cup chopped or diced onion (4 ounces)	=	1 medium onion
1 cup finely chopped onion (7 ounces)	=	1 large onion
1 cup sliced bell pepper — red, green, or yellow (4 ounces)	=	1 small bell pepper

Amount Required in Recipe (Trimmed Weight)	Equals	Amount Whole Vegetable or Nut
1 cup chopped or diced bell pepper (4½ ounces)	=	1 small bell pepper
1 cup finely chopped bell pepper (5 ounces)	=	1 small bell pepper

◆ RECIPE EQUIVALENTS FOR BEANS ◆

Amount Required in Recipe	Equals	Size Can Needed
1 cup cooked beans	=	10 ounces
1½ cups cooked beans	=	16 ounces
1¾ cups cooked beans	=	20 ounces

All canned beans should be drained and rinsed before being used in a recipe.

Soups

❧ LENTIL-APPLE SOUP (E)(V)

The apple adds just the right amount of sweetness to this lentil soup. The easiest way to remove the sand from leeks is to slice them, then place in a strainer and rinse with cold water until clean.

In a 4-quart saucepan, heat the oil over medium-high heat. Add the leek and cook, stirring, until softened. Add the water and apple juice; bring to a boil.

Stir in the lentils, celery, onion, and bay leaf. Return to a boil, reduce the heat, and simmer, uncovered, 45 minutes. Stir in the apple. Simmer, uncovered, 15 minutes longer. Discard the celery, onion, and bay leaf. Stir in the salt.

SERVES 8 TO 10

2 tablespoons vegetable oil
1½ cups thinly sliced leek, white and light green parts only (2 medium leeks), thoroughly rinsed
6 cups water
2 cups unsweetened apple juice
1½ cups lentils, rinsed
3 large ribs celery, with leaves
1 onion, studded with 6 whole cloves
1 bay leaf
1½ cups peeled, chopped apple
1 teaspoon salt

Calories:	200	Total fat:	3.4 g
Protein:	10.5 g	Saturated fat:	1.9 g
Carbohydrates:	33.6 g	Cholesterol:	0
Fiber:	5.3 g	Sodium:	255 mg

∾ MADEIRA–BLACK BEAN SOUP (E)

1 cup black (turtle) beans
1 tablespoon margarine or
  vegetable oil
1 cup sliced leek (white
  and light green parts
  only), thoroughly rinsed
6 cups water
1 cup sliced carrot
1 bay leaf
½ teaspoon dried thyme
¼ cup Madeira
1 teaspoon salt
½ cup milk
  Thinly sliced scallion for
  garnish

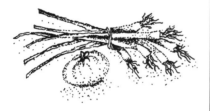

A delicate soup, this is enhanced by the thinly sliced scallion that garnishes it. I use medium-sweet Madeira, but the recipe would be equally good with rainwater (dry) or sweet Madeira.

Place the black beans in a large bowl and add water to a height of at least 2 inches over the beans. Let stand overnight. Drain and rinse until the water runs clear. Set aside.

In a 4-quart saucepan, melt the margarine over medium-high heat. Add the leek and cook, stirring, until softened. Add the water and bring to a boil. Add the beans, carrots, bay leaf, and thyme; return to a boil. Reduce the heat and simmer, covered, 1½ hours. Uncover and simmer 1 hour longer. Stir in the Madeira and salt; simmer 10 minutes longer.

Place the soup in a blender container or the workbowl of a food processor fitted with a steel blade. Cover and process until smooth (you will have to do this in 3 or 4 batches). Return all the soup to the pot and stir in the milk. Simmer 5 minutes, or until heated through.

Serve topped with the scallion.

SERVES 6 TO 8

Calories:	170	*Total fat:*	3.1 g
Protein:	8.2 g	*Saturated fat:*	0.9 g
Carbohydrates:	25.4 g	*Cholesterol:*	2.8 mg
Fiber:	2.3 g	*Sodium:*	403 mg

✍ SWEET POTATO–RED LENTIL SOUP (E)(Q)(V)

This is one of my "you-can-stand-a-spoon-in-it" soups. If at the end of the cooking time you find that it's too thick, stir in as much boiling water as necessary to create the texture you desire. (Be sure to let the soup simmer 3 to 5 minutes after adding the water, so that the flavors blend.) If you want an excellent side dish, puree the soup in a blender or food processor.

1 tablespoon vegetable oil
½ cup chopped onion
3 cups water
2 cups orange juice
2 tablespoons firmly packed light or dark brown sugar
¾ teaspoon salt
½ teaspoon ground ginger
¼ teaspoon ground nutmeg
⅛ teaspoon ground cinnamon
⅛ teaspoon ground red pepper
2½ cups peeled, cubed sweet potato (1-inch pieces)
¾ cup red lentils

In a 3-quart saucepan, heat the oil over medium-high heat. Add the onion and cook, stirring, until soft. Add the water, orange juice, brown sugar, salt, ginger, nutmeg, cinnamon, and red pepper; bring to a boil. Add the sweet potato and lentils; cook, uncovered, over medium heat, 20 minutes, stirring occasionally. Reduce the heat and, stirring frequently, simmer 10 minutes longer, or until the lentils dissolve and the sweet potatoes are soft.

SERVES 4 TO 6

Calories:	287	*Total fat:*	3.6 g
Protein:	10.7 g	*Saturated fat:*	0.7 g
Carbohydrates:	56.2 g	*Cholesterol:*	0
Fiber:	6.5 g	*Sodium:*	441 mg

❧ VEGETABLE-BEAN SOUP (V)

⅓ cup dried small lima
  beans, rinsed
¾ cup (1 ounce) dried
  mushrooms
8 cups water
1 cup chopped onion
1 cup diced carrot
1 cup diced celery
1 cup diced parsnip
1 bay leaf
¼ teaspoon dried thyme
⅓ cup red lentils, rinsed
1 teaspoon salt

I use shiitake mushrooms for this soup. If you can find (or afford) dried Polish or Italian mushrooms, the flavor will be even better. In any case, be sure to use dried mushrooms, not fresh. I dice all the vegetables into ¼-inch pieces.

Place the lima beans in a medium bowl and cover with plenty of water. Let stand overnight, then drain and set aside. Or use the quick-soak method: place the beans in a 1-quart saucepan and fill the pan halfway with water. Bring to a boil and boil 2 minutes; remove from the heat and let stand, covered, 1 hour. Drain and set aside.

Rinse the mushrooms quickly in cold water to remove any grit. Place the mushrooms and the 8 cups of water in a 4-quart saucepan. Bring to a boil and cook 2 minutes, or until the mushrooms are softened. Remove the mushrooms to a cutting board. Discard the tough stems and chop the mushrooms; return to the pan.

Add the onion, carrot, celery, parsnip, bay leaf, thyme, and soaked lima beans. Bring to a boil; reduce the heat and simmer, uncovered, 45 minutes. Stir in the red lentils and simmer, uncovered, 40 minutes longer, stirring occasionally. Stir in the salt and simmer 5 minutes longer.

SERVES 6

Calories:	113	Total fat:	0.5 g
Protein:	6.2 g	Saturated fat:	0
Carbohydrates:	22.4 g	Cholesterol:	0
Fiber:	7.2 g	Sodium:	384 mg

THREE-BEAN SOUP (V)

I puree the parsnip in this soup to add a little body to the broth. You can choose different bean combinations, if you like, but make sure that the soaking and cooking times are similar to those of the beans called for.

In a 6-quart saucepan or stockpot, bring 2 of the 4 quarts of water to a boil. Add the Anasazi, lima, and cow beans, and boil 2 minutes. Remove from the heat, cover, and let stand 1 hour, stirring once, after 30 minutes. Drain the beans and rinse; set aside.

In a 6-quart saucepan or stockpot, heat the oil. Add the leek and cook, stirring, until softened. Stir in the remaining 2 quarts water and bring to a boil. Add the soaked beans, carrot, parsnips, and bay leaves. Return to a boil, reduce the heat, and simmer, covered, 1 hour, or until the beans are soft. Discard the bay leaves.

Remove the parsnips from the pan and place in a blender container along with 2 cups of liquid from the soup. Cover and process on high until smooth. Return to the pan, along with the parsley, garlic, lemon

4 quarts water, divided
¾ cup Anasazi beans
⅔ cup dried lima beans
⅓ cup cow (field) beans
2 tablespoons vegetable oil
1 cup sliced leek (white and light green parts only), thoroughly rinsed
2 cups sliced carrot
2 whole parsnips
2 bay leaves
½ cup chopped parsley
2 large cloves garlic, minced
2 teaspoons fresh lemon juice
1 teaspoon celery salt
½ teaspoon salt
½ teaspoon Red Hot sauce

juice, celery salt, salt, and Red Hot sauce.

Let stand 5 minutes to let the flavors meld. Stir before serving.

SERVES 10 TO 12

Calories:	170	*Total fat:*	3.3 g
Protein:	8.0 g	*Saturated fat:*	0.7 g
Carbohydrates:	28.6 g	*Cholesterol:*	0
Fiber:	8.9 g	*Sodium:*	320 mg

❧ SEVEN-BEAN SOUP (V)

4 quarts water, divided
½ cup kidney beans
⅓ cup pinto beans
⅓ cup dried baby lima beans
⅓ cup black-eyed peas
¼ cup small white beans
¼ cup black (turtle) beans
2 tablespoons olive oil
1½ cups chopped onion
½ cup green split peas
2 cups diced turnip
2 cups chopped cabbage
1½ cups sliced carrot
¼ cup chopped fresh dill *or* 2 teaspoons dillweed
1½ teaspoons salt
¼ teaspoon pepper

This soup, thickened by split peas, is very filling and delicious. The lima beans are the dried type, not fresh or frozen.

In a 6-quart saucepan or stockpot, bring 2 of the 4 quarts of water to a boil. Add the kidney, pinto, and lima beans, the black-eyed peas, and the white and black beans to the pan. Return to a boil and boil 2 minutes. Remove from the heat, cover, and let stand 1 hour. Drain the beans, then rinse; set aside.

In a 6-quart saucepan or stockpot, heat the oil over medium-high heat. Add the onion and sauté until softened. Add the remaining 2 quarts water and bring to a boil. Add the split peas and the soaked beans; return to a boil, reduce the heat, and simmer 15 minutes. Add the turnip, cabbage, and carrot. Bring to a boil. Reduce the heat and simmer, covered, 50 minutes. Add the dill and simmer 10 minutes longer. Stir in the salt and pepper.

SERVES 10 TO 12

Calories:	185	Total fat:	3.3 g
Protein:	10.0 g	Saturated fat:	0.5 g
Carbohydrates:	30.3 g	Cholesterol:	0
Fiber:	9.7 g	Sodium:	351 mg

ITALIAN LENTIL-VEGETABLE SOUP (V)

A combination of vegetables and flavors that I associate with Italian cooking — zucchini, escarole, garlic, and oregano — adds a special flavor to this vegetable soup. If you can't find escarole, you can use Swiss chard or spinach.

In a 6-quart saucepan or stockpot, bring the water to a boil over high heat. Add the lentils, onion, carrot, celery, garlic, bay leaf, and oregano; return to a boil. Reduce the heat and simmer, covered, 1 hour. Add the squash, escarole, zucchini, salt, and pepper; simmer, uncovered, 30 minutes longer. Remove the garlic cloves and bay leaf before serving.

6 cups water
1 cup lentils, rinsed
1 cup chopped onion
1 cup sliced carrot
1 cup sliced celery
2 large cloves garlic
1 bay leaf
¼ teaspoon dried oregano
2 cups sliced yellow squash
2 cups firmly packed, coarsely chopped escarole
1 cup sliced zucchini
1 teaspoon salt
⅛ teaspoon pepper

SERVES 6 TO 8

Calories:	133	Total fat:	0.6 g
Protein:	9.5 g	Saturated fat:	0.1 g
Carbohydrates:	25.4 g	Cholesterol:	0
Fiber:	6.5 g	Sodium:	396 mg

❧ GREEN LENTIL AND VEGETABLE SOUP (V)

2 tablespoons olive oil
1½ cups sliced leek (white
   and light green parts
   only), thoroughly rinsed
1 cup sliced mushrooms
¼ cup finely chopped green
   bell pepper
5 cups water
½ cup green lentils, rinsed
½ teaspoon dried basil
¾ cup diced rutabaga (yellow turnip)
½ cup sliced carrot
½ cup sliced celery
1 cup diced new red po-
   tato
½ cup fresh or frozen cut
   green beans
½ cup diced zucchini
¼ cup chopped parsley
1 clove garlic, minced
1¼ teaspoons salt
¼ teaspoon pepper

Green lentils are very similar to the more common brown lentils, but they have a slightly more delicate flavor. If you cannot find them, substituting brown lentils will be perfectly fine.

Heat the oil in a 4-quart saucepan over medium-high heat. Add the leek, mushrooms, and bell pepper, and sauté, stirring, until softened. Add the water and bring to a boil. Add the lentils and basil. Return to a boil; reduce the heat and simmer, covered, 30 minutes. Add the rutabaga, carrot, and celery; return to a boil and simmer, covered, 20 minutes. Add the potato and simmer, covered, 10 minutes. Add the green beans, zucchini, parsley, garlic, salt, and pepper. Return to a boil and simmer 5 minutes longer, or until the vegetables are tender.

SERVES 8

Calories:	103		*Total fat:*	3.7 g
Protein:	4.5 g		*Saturated fat:*	0.5 g
Carbohydrates:	14.1 g		*Cholesterol:*	0
Fiber:	3.3 g		*Sodium:*	360 mg

❧ DAL (E)(Q)(V)

Dal, although technically a soup, is served as a side dish in Indian cooking. Red lentils are very different from the brown or green lentils you find in the supermarket — they cook much faster, for one thing — and you cannot substitute brown or green for the red variety. You can buy red lentils at health food stores. Garam masala is a blend of ground spices similar to curry powder. Ghee, clarified butter, can be prepared as follows: bring unsalted butter to a boil and simmer until the milk solids in the bottom of the pan turn brown. Let the solids settle and then pour the clear liquid (the ghee) into a container. Discard the milk solids. The nutritional values that follow the recipe have been calculated on the assumption that the recipe was prepared using ghee; if you've used vegetable oil, there is no cholesterol and practically no saturated fat.

1 cup red lentils
2 tablespoons ghee or vegetable oil
½ cup thinly sliced onion
2 cloves garlic, minced
1 teaspoon grated fresh
  ginger *or* ½ teaspoon
  ground ginger
1 teaspoon *garam masala*
  or curry powder
¼ teaspoon ground tur-
  meric
¼ teaspoon ground cumin
3½ cups water
1 tablespoon fresh lemon
  juice
¾ teaspoon salt

Wash the red lentils, discarding any that are floating, and drain well. Set aside.

In a 2-quart saucepan, heat the ghee over medium heat. Add the onion, garlic, and ginger, and cook until the onions are golden. Stir in the *garam masala*, turmeric, and cumin, and cook until absorbed. Add the water and bring to a boil. Add the drained lentils and return to a boil. Reduce the heat and simmer, covered, 30 minutes, stirring once or twice. Stir in the lemon juice and salt; simmer 5 minutes longer.

SERVES 4 AS A SOUP,
6 AS A SIDE DISH

Calories:	212	*Total fat:*	6.4 g
Protein:	11.8 g	*Saturated fat:*	3.7 g
Carbohydrates:	30.3 g	*Cholesterol:*	15.7 mg
Fiber:	6.0 g	*Sodium:*	452 mg

OLD-FASHIONED SPLIT PEA SOUP (E)(V)

10 cups water
1½ cups green split peas
6 ribs celery
4 large carrots
2 whole onions
1 parsnip
6 sprigs celery leaves
1 bay leaf
1 teaspoon celery salt
½ teaspoon poultry seasoning

This is the very thick (and I mean very thick) type of pea soup that my grandmother used to make. If your grandmother made a thinner soup, stir in an extra cup or two of boiling water 30 minutes before the soup is ready. To turn this soup into a one-course meal, stir in some diced tofu toward the end of the cooking time. If you don't have any poultry seasoning, just omit it, or substitute a pinch each of ground sage, thyme, and marjoram.

Bring the water to a boil over high heat in an 8-quart saucepan or stockpot. Add the split peas, celery, carrots, onions, parsnip, celery leaves, bay leaf, celery salt, and poultry seasoning to the water. Return to a boil. Reduce the heat and simmer, uncovered, 2 hours. Remove the carrots from the pan and set aside. Discard the bay leaf. Remove the remaining vegetables from the soup and place in a large strainer. Press the liquid out of the vegetables and return the liquid to the soup. Chop the carrots and return to the soup.

SERVES 8 TO 10

Calories:	147	*Total fat:*	0.5 g
Protein:	9.3 g	*Saturated fat:*	0.1 g
Carbohydrates:	28.0 g	*Cholesterol:*	0
Fiber:	6.9 g	*Sodium:*	305 mg

~ CANADIAN PEA SOUP (V)

I find this a nice thin pea soup. Bouillon cubes or packets are optional, depending on your tastes. If you choose to omit the bouillon, you may want to add salt.

In a 4-quart saucepan, bring the water to a boil over high heat. Add the onion and split peas, and return to a boil. Reduce the heat and simmer, covered, 1 hour.

Add the carrot, celery, dill, barley, bouillon cubes (if using), and pepper. Simmer, uncovered, 1 hour longer, or until the peas have dissolved.

8 cups water
1½ cups chopped onion
¾ cup yellow split peas
1½ cups diced carrot
1½ cups chopped celery
¼ cup chopped fresh dill
3 tablespoons pearl barley
¾ ounce vegetable (or chicken) bouillon cubes or packets (optional)
¼ teaspoon pepper

SERVES 8 TO 10

Calories:	103	*Total fat:*	0.6 g
Protein:	5.8 g	*Saturated fat:*	0
Carbohydrates:	20.0 g	*Cholesterol:*	0
Fiber:	4.5 g	*Sodium:*	370 mg

∿ MUSHROOM-BARLEY SOUP (V)

1 cup boiling water
¾ cup (1 ounce) dried
   mushrooms
1 tablespoon vegetable oil
2 cups sliced fresh mush-
   rooms
1½ cups chopped onion
2 cups vegetable (or beef)
   broth
5 cups water
1 cup chopped carrot
1 cup chopped celery
3 tablespoons pearl barley,
   rinsed
1 bay leaf
⅛ teaspoon dried thyme
⅛ teaspoon pepper

Polish mushrooms are best for this soup because they are extremely flavorful — but they are also extremely expensive. Use dried shiitake mushrooms if you don't have Polish ones.

In a small bowl, combine the boiling water and dried mushrooms. Let stand 10 minutes, or until the mushrooms are softened. Chop the mushrooms and set aside, reserving the soaking water.

In a 4-quart saucepan, heat the oil over medium-high heat. Add the fresh mushrooms and onion, and cook, stirring, until softened. Add the broth, water, carrot, celery, and the reserved dried mushrooms and soaking water. Bring to a boil. Reduce the heat and simmer, uncovered, 45 minutes. Add the barley, bay leaf, thyme, and pepper. Simmer 40 minutes longer. Discard the bay leaf before serving.

SERVES 8

Calories:	71	*Total fat:*	2.3 g
Protein:	2.0 g	*Saturated fat:*	0.3 g
Carbohydrates:	12.1 g	*Cholesterol:*	0
Fiber:	3.5 g	*Sodium:*	435 mg

~ MINESTRONE (V)

This is the kind of soup in which you can put everything but the kitchen sink. I use Italian paste, which is tomato paste with Italian spices and is available in some supermarkets. You can use plain tomato paste and add some oregano to the soup instead. For a quick version, omit the dried beans. Start by sautéing the onion and cabbage; then add 8 cups water, bring to a boil, add the remaining vegetables, and cook 30 minutes. Add ½ cup canned kidney beans and simmer 10 minutes longer.

10 cups water, divided
½ cup dried Roman (cranberry) or kidney beans, rinsed
2 tablespoons olive oil
1 cup chopped onion
2 cups shredded cabbage
1 can (14½ ounces) whole peeled tomatoes
1 cup sliced carrot
1 cup sliced celery
1 can (6 ounces) Italian paste
1 cup sliced zucchini
½ cup *ditalini* or pasta of choice
1¼ teaspoons salt
¼ teaspoon pepper
⅓ cup grated Parmesan (optional)

Bring 2 of the 10 cups of water to a boil and add the beans. Boil 2 minutes. Let stand 1 hour, then drain and rinse. Set aside.

In a 12-quart stockpot, heat the oil. Stir in the onion and sauté until soft. Add the cabbage and sauté until wilted. Add the remaining 8 cups of water and bring to a boil. Stir in the drained beans and simmer, covered, 1½ to 2 hours, or until the beans are soft. (The beans must be completely cooked at this point because they will cease to soften once you add the tomatoes.)

Stir in the tomatoes, breaking them up with the back of a spoon, and then add the carrot, celery, and Italian paste. Return to a boil, then simmer, covered, 30 minutes. Stir in the zucchini, *ditalini*, salt, and pepper. Return to a boil and simmer, uncovered, 12 minutes, or until the *ditalini* are tender. Stir in the Parmesan (if using) and serve.

SERVES 10 TO 12

Calories:	160	*Total fat:*	4.6 g
Protein:	6.9 g	*Saturated fat:*	1.0 g
Carbohydrates:	24.3 g	*Cholesterol:*	2.0 mg
Fiber:	5.7 g	*Sodium:*	230 mg

◞ FRESH VEGETABLE SOUP   (E)(Q)(V)

1 tablespoon vegetable oil
1 cup chopped onion
4 cups vegetable (or chicken) broth
1 can (14½ ounces) whole peeled tomatoes, undrained
1 can (6 ounces) tomato paste
1 cup dry white wine
⅓ cup chopped parsley
1 teaspoon dried oregano
¼ teaspoon pepper
1 bay leaf
1½ cups sliced carrot
1½ cups sliced celery
1 cup green beans, cut into 1-inch pieces
1 cup chopped zucchini
1 can (7 ounces) corn kernels, undrained

Although most soups start with fresh vegetables, this one cooks for only a short time, allowing the vegetables to retain their individual characteristics.

In a 4-quart saucepan, heat the oil over medium-high heat. Add the onion and cook, stirring, until softened. Add the broth and then the tomatoes, breaking them up with the back of a spoon. Add the tomato paste, wine, parsley, oregano, pepper, and bay leaf. Add the carrot and celery, and bring to a boil. Lower the heat and simmer, uncovered, 15 minutes. Stir in the green beans, zucchini, and corn; simmer 10 minutes longer, or until the vegetables are tender-crisp. Remove the bay leaf before serving.

SERVES 10

Calories:	76	*Total fat:*	2.3 g
Protein:	2.8 g	*Saturated fat:*	0
Carbohydrates:	13.3 g	*Cholesterol:*	0
Fiber:	3.1 g	*Sodium:*	504 mg

❧ SHREDDED CABBAGE SOUP (E)(Q)(V)

Your guests may never guess there are lime and ginger in this soup; they'll just know it's delicious! It's important to the texture of the soup that the cabbage be soft before you add the remaining ingredients to the pot.

In a 6-quart saucepan or stockpot, heat the oil over medium-high heat. Add the cabbage and onion, and cook, stirring, until the vegetables are softened. Stir in the ginger and then the tomatoes in puree, breaking up the tomatoes with the back of a spoon. Add the water, sugar, lime juice, salt, and pepper. Bring to a boil. Reduce the heat and simmer 20 minutes.

SERVES 8 TO 10

3 tablespoons vegetable oil
6 cups shredded cabbage
1½ cups chopped onion
½ teaspoon ground ginger
2 cans (16 ounces each) whole tomatoes in thick puree
4 cups water
¼ cup firmly packed light or dark brown sugar
1 tablespoon fresh lime or lemon juice
1 teaspoon salt
¼ teaspoon pepper

Calories:	115	*Total fat:*	5.5 g
Protein:	2.0 g	*Saturated fat:*	0.7 g
Carbohydrates:	16.5 g	*Cholesterol:*	0
Fiber:	2.9 g	*Sodium:*	463 mg

❧ ONION SOUP (V)

2 tablespoons olive oil
4 cups sliced Spanish or
  Bermuda onion
6 cups vegetable (or beef)
  broth
⅔ cup dry red wine
2 tablespoons dry ver-
  mouth
2 tablespoons brandy
1 bay leaf
¼ teaspoon dried thyme

This onion soup has a full, rich flavor, which is just wonderful as is. If you'd like to make French onion soup, put the soup into individual ovenproof bowls, float a slice of stale French bread or a melba round in each bowl, and top with plenty of grated Gruyère or Swiss cheese and sprinkle with grated Parmesan. Place under a preheated broiler and cook until the cheese is melted and slightly browned on top.

In a 4-quart saucepan, heat the oil over medium heat. Add the onion and cook, stirring, until very soft. Stir in the broth, wine, vermouth, and brandy. Add the bay leaf and thyme. Bring to a boil; re duce the heat and simmer 1 hour. Discard the bay leaf before serving.

SERVES 8

Calories:	85	*Total fat:*	3.2 g
Protein:	1.7 g	*Saturated fat:*	0.1 g
Carbohydrates:	6.9 g	*Cholesterol:*	0
Fiber:	0.9 g	*Sodium:*	597 mg

⌾ SWISS CHARD SOUP (E)(V)

I served this very tasty and easy soup to friends, and they couldn't believe that it took me less than 15 minutes to prepare (using canned beans). You can substitute bok choy if Swiss chard isn't available.

In a 3-quart saucepan, heat the oil over medium-high heat. Add the Swiss chard and onion and cook, stirring, until softened. Add the broth, cannellini, and lemon juice. Bring to a boil; reduce the heat and simmer, uncovered, 10 minutes.

SERVES 6

2 tablespoons olive oil
2 cups coarsely chopped Swiss chard
1 cup chopped onion
4 cups vegetable (or chicken) broth
1 cup cooked or canned (drained and rinsed) cannellini
1 tablespoon fresh lemon juice

Calories:	126	*Total fat:*	5.6 g
Protein:	5.4 g	*Saturated fat:*	0.8 g
Carbohydrates:	14.2 g	*Cholesterol:*	0.7 mg
Fiber:	4.8 g	*Sodium:*	570 mg

⌾ HOUSE SPECIAL SOUP (E)(Q)(V)

This soup is reminiscent of those I frequently order at my local Chinese restaurant. You may want to dilute your broth a little, to make the flavor less intrusive.

In a 2-quart saucepan, heat the oil over medium-high heat. Add the cabbage and cook, stirring, until wilted. Stir in the broth, water, *mirin,* and soy sauce, and bring to a boil. Add the tofu, broccoli, snow

1 tablespoon oil
1½ cups coarsely chopped cabbage
2 cups vegetable (or chicken) broth
1 cup water
2 teaspoons *mirin* or dry sherry
1 teaspoon soy sauce
1 cup (6 ounces) diced tofu
1 cup broccoli florets

½ cup chopped snow peas
¼ cup fresh or frozen peas
2 tablespoons sliced scallion

Calories:	71	Total fat:	4.7 g
Protein:	3.2 g	Saturated fat:	0.7 g
Carbohydrates:	4.7 g	Cholesterol:	0.5 mg
Fiber:	2.0 g	Sodium:	524 mg

peas, and peas. Reduce the heat and simmer 5 minutes. Stir in the scallion.

SERVES 4 TO 6

❧ HARIRAH (V)

1½ tablespoons olive oil
2 cups chopped onion
1 cup chopped celery
1 teaspoon ground tur-
   meric
½ teaspoon ground ginger
3 cups vegetable (or
   chicken) broth
3 cups water
⅓ cup dried green or yel-
   low split peas
1 cinnamon stick (2½
   inches)
⅓ cup dried lentils
2 tablespoons pearl barley
1 can (14½ ounces) whole
   peeled tomatoes, un-
   drained
¼ cup thin soup noodles
2 tablespoons fresh lemon
   juice
2 tablespoons chopped
   fresh coriander (cilantro)
⅛ teaspoon pepper

This is probably the only recipe in the book whose title is a palindrome (it's the same spelled backward or forward), but that's not the only unusual element about it. Its unexpected combination of flavors reflects its Moroccan roots, and the tomatoes and lemon juice add a lovely tartness.

Heat the oil in a 4-quart saucepan over medium-high heat. Add the onion and celery, and cook, stirring, until softened. Stir in the turmeric and ginger until absorbed. Add the broth, water, split peas, and cinnamon stick. Bring to a boil, reduce the heat, and simmer 45 minutes. Add the lentils and barley. Simmer, uncovered, 40 minutes. Add the tomatoes, breaking them up with the back of a spoon, and noodles; simmer 10 minutes longer. Stir in the lemon juice, coriander, and pepper. Discard the cinnamon stick before serving.

SERVES 6 TO 8

Calories:	172	Total fat:	4.6 g
Protein:	8.4 g	Saturated fat:	0.7 g
Carbohydrates:	26.3 g	Cholesterol:	0
Fiber:	5.3 g	Sodium:	470 mg

❧ GAZPACHO (E)(Q)(V)

The following recipe does not deliver the same deep-red gazpacho that some others do. If you want a deeper color, add a little tomato juice, but remember that you'll be diluting the flavor, and adjust the seasonings accordingly.

Cut the tomato, cucumber, bell pepper, and onion into chunks. Drop the chunks and the garlic into the container of a blender, through the hole in the lid, with the blender running. Process until smooth. You may have to do this in 2 or 3 batches, in which case just empty each batch into a large bowl and continue pureeing the remaining vegetables. (If you're using a food processor, you can put all the vegetable chunks in the workbowl, fitted with a steel blade, and then process until smooth.) Stir in the vinegar, oil, salt, and Red Hot sauce.

Chill before serving.

SERVES 4

2 large ripe tomatoes
1 large cucumber (about 8 inches long), peeled
½ medium red or green bell pepper
½ small onion
½ clove garlic
1½ tablespoons red wine vinegar
2 teaspoons olive oil
½ teaspoon salt
½ teaspoon Red Hot sauce

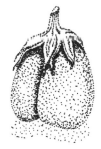

Calories:	43	Total fat:	2.5 g
Protein:	0.9 g	Saturated fat:	0.3 g
Carbohydrates:	5.2 g	Cholesterol:	0
Fiber:	1.4 g	Sodium:	273 mg

❧ GREEN GAZPACHO (E)(Q)(V)

2 large cucumbers (8 to 9 inches long), peeled and cut into pieces
1 medium green bell pepper, cut into pieces
2 large scallions, cut into pieces
6 sprigs fresh coriander (cilantro), large stems discarded
1 small clove garlic, minced
2 teaspoons distilled white vinegar
1 teaspoon olive oil
1 teaspoon fresh lemon juice
1 teaspoon salt
¼ teaspoon Red Hot sauce

This very easy soup is perfect on a hot summer day. Serve it topped with a dollop of unflavored yogurt. If you're not fond of fresh coriander, you can use fresh dill instead. You may have to make this soup in two batches if you have a regular-size blender.

Drop the pieces of cucumber into a blender container, through the hole in the lid, with the blender running. Add the bell pepper, scallions, coriander, and garlic to the running blender.

Add the vinegar, oil, lemon juice, salt, and Red Hot sauce to the blender. Blend until combined.

SERVES 3 TO 4

Calories:	50	*Total fat:*	2.0 g
Protein:	1.6 g	*Saturated fat:*	0.3 g
Carbohydrates:	8.4 g	*Cholesterol:*	0
Fiber:	2.9 g	*Sodium:*	452 mg

❧ VEGETABLE CHOWDER (V)

3 tablespoons vegetable oil
4 cups coarsely shredded cabbage
3 cups sliced leek (white and light green parts only), thoroughly rinsed
5 cups water

You can prepare this soup using frozen lima beans instead of dried, if you like. Stir them in when you add the corn.

In a 4-quart saucepan, heat the oil over medium-high heat. Add the cabbage and leek, and cook, stirring, until wilted. Add the water and bring to a boil. Add the beans and bay leaf, and simmer, covered, 40 minutes.

Add the tomatoes, breaking them up with the back of a spoon, and the sugar, barley, tomato paste, celery salt, and pepper. Bring to a boil; reduce the heat and simmer 20 minutes. Add the corn and simmer 20 minutes longer.

SERVES 10

½ cup dried baby lima beans, rinsed
1 bay leaf
1 can (14½ ounces) whole peeled tomatoes, undrained
¼ cup firmly packed light or dark brown sugar
2 tablespoons pearl barley
2 tablespoons tomato paste
1 teaspoon celery salt
¼ teaspoon pepper
1 can (10 ounces) corn kernels, undrained

Calories:	191	*Total fat:*	5.8 g
Protein:	5.9 g	*Saturated fat:*	0.8 g
Carbohydrates:	32.1 g	*Cholesterol:*	0
Fiber:	6.7 g	*Sodium:*	487 mg

✎ CORN CHOWDER (E)(Q)

I like my chowders on the thin side; if you prefer yours thick, use 3 tablespoons of flour instead of the 2 tablespoons called for below.

In a 2-quart saucepan, bring the broth to a boil over medium-high heat. Add the corn, potato, onion, celery, bay leaf, and thyme. Cook, uncovered, 20 minutes, or until the potatoes are soft.

While the vegetables are cooking, melt the margarine in a 3-quart saucepan. Stir in the flour until absorbed. Using a whisk, stir in the milk and pepper. Cook, stirring, until thickened. Stir in the vegetable-

2 cups vegetable broth
2 cups fresh or frozen corn kernels
1 cup peeled, diced potato (½-inch pieces)
½ cup chopped onion
⅓ cup finely chopped celery
1 bay leaf
¼ teaspoon dried thyme
2 tablespoons margarine
2 tablespoons all-purpose flour
2 cups milk
⅛ teaspoon pepper

broth mixture and cook 5 minutes longer. Discard the bay leaf before serving.

SERVES 6

Calories:	170	*Total fat:*	7.6 g
Protein:	5.7 g	*Saturated fat:*	2.6 g
Carbohydrates:	21.8 g	*Cholesterol:*	11.3 mg
Fiber:	3.5 g	*Sodium:*	384 mg

❧ SHAKER SUCCOTASH SOUP (E)(Q)

---

1 cup vegetable (or chicken) broth
1 cup fresh or frozen baby lima beans
2 tablespoons margarine
¼ cup sliced leek (white and light green parts only), thoroughly rinsed
3 tablespoons all-purpose flour
1 can (5 ounces) evaporated milk
½ cup milk
1 can (15 to 17 ounces) corn kernels, undrained
⅛ teaspoon pepper

A cross between cream of corn soup and succotash, this soup is creamy and delicious.

In a small saucepan, bring the broth to a boil over medium-high heat. Add the lima beans and return to a boil. Reduce the heat and simmer, covered, 10 minutes, or to the desired doneness.

While the beans are cooking, melt the margarine in a 2-quart saucepan over medium-high heat. Add the leek and cook, stirring, until softened. Stir in the flour until absorbed. Add the evaporated and fresh milk, and cook, stirring with a whisk, until smooth. Add the corn (with its liquid) and pepper, and cook, stirring, until the mixture comes to a boil and thickens. Add the beans and broth; cook until heated through.

SERVES 4 TO 6

Calories:	225	Total fat:	6.4 g
Protein:	9.4 g	Saturated fat:	2.7 g
Carbohydrates:	35.2 g	Cholesterol:	14.8 mg
Fiber:	5.7 g	Sodium:	556 mg

‌ CREAM OF CORN SOUP WITH RED PEPPER PUREE (Q)

This corn soup is delicious even on its own, but adding the red pepper puree gives it a Southwest feel. You can use ground red pepper, if you like, to make the contrast between a relatively bland soup and a strongly flavored puree even greater.

For the puree: Cut the bell pepper in half and discard the stem and seeds. Place in a preheated broiler and cook until charred. Turn and broil until charred on the other side. (To microwave: cut the pepper in quarters, discarding the stem and seeds. Place on wax paper in the oven and microwave 2 minutes on high [100 percent] power. Turn the pieces over and microwave 1 minute longer on high power.) Remove the pepper from the oven and let stand in a paper bag 10 minutes. Strip off the skin and place the peeled pepper in a blender container with the broth, tomato paste, garlic, and ground red pepper (if using). Cover and process until smooth. (Makes ⅔ cup puree.) Rinse the blender container.

For the soup: Reserve 1 cup corn kernels; place the remaining corn (with its liquid) in a blender con-

PUREE
- 1 red bell pepper
- ¼ cup vegetable broth or water
- 1 tablespoon tomato paste
- 1 clove garlic, minced
- ⅛ teaspoon ground red pepper (optional)

SOUP
- 2 cans (15 to 17 ounces each) corn kernels, undrained
- 2 cups milk
- 2 tablespoons margarine
- 3 tablespoons all-purpose flour
- ⅛ teaspoon pepper

tainer or the workbowl of a food processor fitted with a steel blade. Cover and process (you may have to do this in more than 1 batch), adding as much of the 2 cups milk as necessary to make a puree. Gradually stir in any remaining milk. In a 3-quart saucepan, melt the margarine over medium-high heat. Stir in the flour until absorbed. Gradually stir in the milk and corn puree. Cook, stirring, until the mixture comes to a boil. Stir in the reserved corn and pepper. Remove from the heat and set aside.

Pour the soup into individual bowls. Drizzle the red pepper puree into the center of each bowl, forming a pinwheel toward the outer edge of the bowl. Using the tip of a knife, swirl to make an attractive pattern.

SERVES 6

Calories:	201	*Total fat:*	7.4 g
Protein:	6.4 g	*Saturated fat:*	2.6 g
Carbohydrates:	31.2 g	*Cholesterol:*	11.0 mg
Fiber:	4.8 g	*Sodium:*	507 mg

✺ CREAM OF ASPARAGUS AND CARROT SOUP (E)(Q)

---

2 teaspoons olive oil
1 cup sliced leek (white and light green parts only), thoroughly rinsed
4 cups vegetable (or chicken) broth

Asparagus and carrots are an unexpected combination, one that provides a delicious, mysterious flavor. The buttermilk adds a lovely tanginess. If you are not fond of buttermilk, use regular milk.

Heat the oil in a 3-quart saucepan over medium-high heat. Add the leek and cook, stirring, until softened. Add the broth, carrot, asparagus, and celery; bring to a boil. Reduce the heat and simmer, uncovered, 30 minutes or until the carrot and asparagus are very soft.

Place the broth and vegetables in a blender container or the workbowl of a food processor fitted with a steel blade. Cover and process until smooth. (You may have to do this in 2 batches.) Return to the pan and stir in the milk, buttermilk, and red pepper. If necessary, reheat.

SERVES 6 TO 8

2 cups sliced carrot
2 cups fresh asparagus pieces
1 cup sliced celery
1 cup milk
¾ cup buttermilk
⅛ teaspoon ground red pepper

Calories:	109	Total fat:	4.2 g
Protein:	5.6 g	Saturated fat:	1.5 g
Carbohydrates:	13.9 g	Cholesterol:	7.0 mg
Fiber:	2.9 g	Sodium:	656 mg

～ POTAGE CRÉCY (E)(Q)

Serve this creamy carrot-potato-leek soup warm, garnished with an extra dollop of yogurt and snipped chives. If you are reheating or warming it, do so slowly, without letting it boil, to prevent any curdling or separation.

In a 4-quart saucepan, heat the oil over medium-high heat. Add the leek and cook, stirring, until softened. Add the broth and water, and bring to a boil. Add the potato and carrot, and cook, simmering, un-

1 tablespoon olive oil
¾ cup sliced leek (white and light green parts only), thoroughly rinsed
2 cups vegetable broth
2 cups water
2 cups peeled, cubed potato
2 cups sliced carrot
2 cups milk
½ cup unflavored yogurt
½ teaspoon salt
¼ teaspoon ground nutmeg

¼ teaspoon ground red
  pepper

covered, 20 minutes, or until they
are soft.

Puree the soup with the milk
and yogurt in several batches,
using a blender or a food processor fitted with a steel blade. Stir in
the salt, nutmeg, and red pepper.
If necessary, reheat the soup before serving.

SERVES 6 TO 8

Calories:	136	*Total fat:*	5.7 g
Protein:	5.2 g	*Saturated fat:*	2.3 g
Carbohydrates:	16.6 g	*Cholesterol:*	12.5 mg
Fiber:	1.0 g	*Sodium:*	342 mg

❧ CURRIED SQUASH SOUP (E)(Q)

1 tablespoon vegetable oil
¾ cup chopped onion
¾ cup chopped celery
¾ cup chopped carrot
1 tablespoon curry powder
1 teaspoon ground corian-
  der
3 cups water
2 cups vegetable (or
  chicken) broth
3 cups cubed butternut
  squash
1 cup peeled, cubed sweet
  potato
1 cup peeled, cubed apple
1 bay leaf
1 teaspoon salt
¼ teaspoon poultry seasoning

The buttermilk and lemon juice add a tartness to this soup that complements the curry quite well. For a special touch, drizzle and swirl a little extra buttermilk over the top of each bowlful, or serve with a spoonful of unflavored yogurt.

In a 6-quart saucepan or stockpot, heat the oil over medium-high
heat. Add the onion, celery, and carrot. Cook, stirring, until the onion
is softened. Stir in the curry powder
and ground coriander until absorbed. Add the water, broth,
squash, potato, apple, bay leaf, salt,
poultry seasoning, and red pepper
(if using). Bring to a boil. Reduce
the heat and simmer, covered, 30

minutes, or until the vegetables are soft. Discard the bay leaf.

Place a few cups of the soup in a blender container or the workbowl of a food processor fitted with a steel blade. Cover and process until smooth. Repeat until all the soup is pureed. Return to the pan and stir in the buttermilk and lemon juice. Reheat if necessary.

1 cup buttermilk
1 tablespoon fresh lemon juice
⅛ teaspoon ground red pepper (optional)

SERVES 10

Calories:	82	Total fat:	2.3 g
Protein:	2.1 g	Saturated fat:	0.5 g
Carbohydrates:	14.3 g	Cholesterol:	1.0 mg
Fiber:	3.0 g	Sodium:	421 mg

❧ PUREED WHITE BEAN AND PARSNIP SOUP (E)

This smooth, rich soup is as wonderful served cold as it is warmed. You can thin it to your taste by adding milk.

In a 6-quart saucepan or stockpot, bring the broth to a boil over medium-high heat. Add the parsnip and 3 tablespoons of the chives. Return to a boil, lower the heat, and simmer 15 minutes, uncovered. Add the beans and simmer 15 minutes longer. Remove from the heat; stir in the nutmeg and red pepper.

Place 3 cups of the soup in a blender container or the workbowl of a food processor fitted with a steel blade. Cover and process until

6 cups vegetable (or chicken broth)
4 cups peeled, cubed parsnip
¼ cup snipped chives, divided
2 cups cooked or canned (drained and rinsed) small white beans
¼ teaspoon ground nutmeg
⅛ teaspoon ground red pepper
1 cup unflavored yogurt
1 cup milk (optional)

smooth. Pour into a bowl or separate saucepan. Repeat until the soup is completely pureed. Stir in the yogurt and as much of the milk (if using) as necessary to obtain the desired consistency. Reheat if serving warm. Be careful not to boil the soup or the yogurt will curdle. Top with the remaining chives.

SERVES 8 TO 10

Calories:	166	*Total fat:*	3.3 g
Protein:	7.8 g	*Saturated fat:*	1.5 g
Carbohydrates:	27.4 g	*Cholesterol:*	8.6 mg
Fiber:	6.6 g	*Sodium:*	674 mg

Main Dishes

❧ JAMBALAYA RICE AND BEANS (E)(V)

Although traditionally jambalaya is chock-full of meaty items such as ham, shrimp, and sausage, this vegetarian version is as tasty as can be.

Heat the oil in a 3-quart saucepan over medium-high heat. Add the onion, celery, bell pepper, and garlic. Cook, stirring, until the vegetables are soft. Add the tomatoes and break up with the back of a spoon. Stir in the broth, thyme, and red pepper. Bring to a boil. Stir in the rice. Return to a boil, reduce the heat, and simmer, covered, 15 minutes. Stir in the beans, parsley, and vinegar. Return to a simmer and cook, covered, 15 minutes longer. Stir in the salt.

SERVES 4 TO 6

2 tablespoons oil
1 cup chopped onion
½ cup finely chopped celery
½ cup chopped green bell pepper
3 cloves garlic, minced
1 can (14½ ounces) whole peeled tomatoes in thick puree, undrained
1¼ cups vegetable (or chicken) broth
½ teaspoon dried thyme
¼ teaspoon ground red pepper
¾ cup long-grain white rice
1½ cups cooked or canned (drained and rinsed) red kidney beans
2 tablespoons chopped parsley
2 teaspoons distilled white vinegar
½ teaspoon salt

Calories:	320	*Total fat:*	7.9 g
Protein:	11.0 g	*Saturated fat:*	1.2 g
Carbohydrates:	52.9 g	*Cholesterol:*	0.2 mg
Fiber:	8.7 g	*Sodium:*	502 mg

☙ MOROCCAN STEW WITH COUSCOUS (V)

STEW

- 1 tablespoon vegetable oil
- 1½ cups chopped onion
- 2 cloves garlic, minced
- ¾ teaspoon salt, divided
- 1 teaspoon ground cinnamon
- ½ teaspoon ground ginger
- ½ teaspoon ground turmeric
- ¼ teaspoon ground nutmeg
- ¼ teaspoon ground red pepper
- 2 cups water
- 3 whole cloves
- 2 cups sliced carrot
- 2 cups cubed butternut squash
- 2 cups cooked or canned (drained and rinsed) chickpeas
- 1½ cups cubed sweet potato
- ½ cup raisins
- ⅓ cup chopped dried apricots
- 3 tablespoons firmly packed light or dark brown sugar

COUSCOUS

- 1 cup couscous
- 1¾ cups boiling water

TOPPING

- ⅓ cup chopped blanched almonds

This stew is sweet, tart, and spicy all at once; it's a perfect partner for basically bland couscous.

For the stew: In a 4-quart saucepan, heat the oil over medium-high heat. Add the onion and garlic, and cook, stirring, until softened. Add ½ teaspoon of the salt and all the cinnamon, ginger, turmeric, nutmeg, and red pepper, stirring until absorbed. Add the water and cloves; bring to a boil. Add the carrot, squash, chickpeas, sweet potato, raisins, apricots, and brown sugar, and return to a boil. Reduce the heat and simmer uncovered, stirring occasionally, 40 to 45 minutes, or until the sweet potato is tender.

For the couscous: During the last 5 minutes of cooking the stew, place the couscous in a medium-sized bowl. Add the boiling water and remaining ¼ teaspoon salt. Let stand 5 minutes, then fluff with a fork.

Serve the stew over the couscous and top with the chopped almonds.

SERVES 4

Calories:	575	Total fat:	16.3 g
Protein:	20.0 g	Saturated fat:	2.0 g
Carbohydrates:	126.7 g	Cholesterol:	0
Fiber:	24.7 g	Sodium:	447 mg

~ MEDITERRANEAN FAVA BEANS AND BULGUR (V)

Moist and almost stewlike, this dish has a fresh, tart flavor. For a drier consistency, you can leave the cover off for the last 10 minutes of cooking. I like the flavor with 2 tablespoons of lemon juice, but you may want to start with 1 tablespoon and go from there. If you're not partial to mint, use chopped parsley instead.

In a 3-quart saucepan, heat the oil over medium-high heat. Add the eggplant, onion, bell pepper, and garlic. Cook, stirring, until softened. Add the water and bring to a boil; stir in the bulgur, salt, and pepper. Return to a boil, reduce the heat, and simmer, covered, 15 minutes. Stir in the beans, tomato, mint, and lemon juice; simmer, covered, 10 minutes longer.

SERVES 4

2 tablespoons olive oil
2 cups chopped eggplant (¾-inch pieces)
1 cup chopped onion
½ cup chopped red bell pepper
2 cloves garlic
1 cup water
½ cup bulgur or cracked wheat
½ teaspoon salt
¼ teaspoon pepper
2 cups cooked or canned (drained and rinsed) fava beans
1 cup chopped tomato
¼ cup chopped fresh mint
1 to 2 tablespoons fresh lemon juice

Calories:	285	Total fat:	7.8 g
Protein:	11.5 g	Saturated fat:	1.9 g
Carbohydrates:	45.1 g	Cholesterol:	0
Fiber:	14.3 g	Sodium:	326 mg

❧ BARLEY AND BEANS WITH SPINACH, MUSHROOMS, AND PIGNOLI (V)

1 bunch (about ½ pound)
    fresh spinach
¼ cup pignoli (pine nuts)
2 tablespoons olive oil
1½ cups sliced fresh mush-
    rooms
¾ cup chopped onion
3 cups cooked pearl barley
1¾ cups cooked or canned
    (drained and rinsed) can-
    nellini
1 tablespoon fresh lemon
    juice
½ teaspoon dried thyme
¼ teaspoon salt
¼ teaspoon pepper

Although I use cultivated mush-rooms in this recipe, for an even fuller flavor you can substitute fresh shiitake mushrooms for part or all of the 1½ cups called for.

Thoroughly rinse and coarsely chop the spinach, discarding the tough stems; set aside.

Place the pignoli in a dry skillet and cook over medium heat, shaking the skillet, until the nuts are browned. Remove from the skillet and set aside.

Heat the oil in a large skillet over medium-high heat. Add the mushrooms and onion, and cook until softened. Stir in the spinach and cook until just wilted (there will be some liquid in the bottom of the pan). Stir in the barley, beans, lemon juice, thyme, salt, and pepper. Cook until heated through. Stir in the toasted pignoli.

SERVES 4 TO 6

Calories:	433	*Total fat:*	18.4 g
Protein:	16.3 g	*Saturated fat:*	2.3 g
Carbohydrates:	57.6 g	*Cholesterol:*	0
Fiber:	11.3 g	*Sodium:*	242 mg

✎ BAKED CHICKPEAS WITH EGGPLANT AND TOMATOES (E)(V)

I like to serve this dish with pasta on the side to complete the protein; the garlicky broth from the vegetables makes a perfect sauce for the pasta. To make this a vegan dish, omit the Parmesan.

Preheat oven to 350° F.

In a small bowl, stir together the oil, garlic, basil, thyme, ¼ teaspoon salt, and pepper; set aside.

Place the eggplant slices in a colander and sprinkle liberally with additional salt. Let drain 30 minutes; rinse and pat dry. In a 9-inch pie plate, alternate the eggplant and tomato slices in a circle around the outer edge of the dish. Place the chickpeas in the center. Drizzle the oil mixture over the vegetables and chickpeas. Sprinkle with the Parmesan (if using). Bake 1 hour, or until the cheese is browned and the vegetables soft.

SERVES 4

2 tablespoons olive oil
1 clove garlic, minced
¼ teaspoon dried basil
¼ teaspoon dried thyme
¼ teaspoon salt + additional salt
⅛ teaspoon pepper
1 medium eggplant, cut into ¼-inch slices
2 large tomatoes, sliced
1½ cups cooked or canned (drained and rinsed) chickpeas
⅓ cup grated Parmesan (optional)

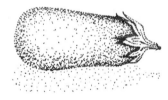

Calories:	312	Total fat:	14.2 g
Protein:	11.0 g	Saturated fat:	2.9 g
Carbohydrates:	51.7 g	Cholesterol:	5.2 mg
Fiber:	10.7 g	Sodium:	351 mg

❧ AUTUMN QUINOA AND BUTTER BEANS (Q)(V)

½ cup quinoa
2 tablespoons margarine
¾ cup finely chopped onion
1 tablespoon minced fresh ginger
¾ cup orange juice
⅔ cup water
2 tablespoons honey
½ teaspoon salt
¼ teaspoon ground coriander
¼ teaspoon ground cardamom
⅛ teaspoon ground nutmeg
1 cup diced sweet potato (½-inch pieces)
1 cup diced butternut squash (½-inch pieces)
1½ cups cooked or canned (drained and rinsed) butter beans
¼ cup chopped cranberries

I use the tartness of the cranberries to contrast with the slight sweetness of the other ingredients. They also add a lovely speck of color.

Thoroughly rinse the quinoa by placing it in a large bowl and filling the bowl with cold water. Drain the quinoa and repeat the rinsing and draining 4 more times; set aside.

Melt the margarine in a 2-quart saucepan over medium-high heat. Add the onion and ginger, and cook, stirring, until the onion is softened. Stir in the orange juice, water, honey, salt, coriander, cardamom, and nutmeg; bring to a boil. Stir in the sweet potato and squash; bring to a boil. Cook, uncovered, 7 minutes. Stir in the butter beans and quinoa, and return to a boil. Reduce the heat and simmer, covered, 15 minutes. Stir in the cranberries; simmer, covered, 5 minutes longer.

SERVES 4

Calories:	345	*Total fat:*	6.7 g
Protein:	10.8 g	*Saturated fat:*	1.3 g
Carbohydrates:	56.0 g	*Cholesterol:*	0
Fiber:	8.8 g	*Sodium:*	392 mg

❧ BARLEY-STUFFED GREEN PEPPERS (E)(V)

I use any jarred sauce for this, but for a special treat, try the marinara sauce on page 133; just omit the beans.

Preheat the oven to 350° F.

Heat the oil in a large skillet. Add the mushrooms and onions, and cook, stirring, until the onions are browned. Stir in the barley, parsley, thyme, and pepper. Stir in the cheese; set aside.

Rinse the bell peppers. Cut off the tops; remove and discard the seeds and pith. If necessary, make a thin slice on the bottom of each pepper to balance. Spoon one quarter of the barley mixture into each pepper. Stand the peppers upright in a baking dish just large enough to accommodate them. Pour the sauce into the baking dish. Bake 30 minutes, or until the peppers are tender.

SERVES 4

1 tablespoon vegetable oil
2 cups chopped mushrooms
1 cup chopped onion
1½ cups cooked pearl barley
2 tablespoons chopped parsley
¼ teaspoon dried thyme
¼ teaspoon pepper
1 cup grated Monterey Jack (optional)
4 medium green bell peppers
1 cup marinara sauce

Calories:	335	*Total fat:*	16.2 g
Protein:	12.3 g	*Saturated fat:*	6.5 g
Carbohydrates:	38.7 g	*Cholesterol:*	26.3 mg
Fiber:	6.0 g	*Sodium:*	472 mg

✌ RED LENTILS AND CUCUMBER RICE (E)(Q)(V)

RICE
2½ cups water, divided
¾ cup white rice
½ cup peeled, seeded, and chopped cucumber
2 tablespoons sliced scallion
2 tablespoons chopped fresh coriander (cilantro)
1 tablespoon margarine (optional)
¾ teaspoon salt, divided

LENTILS
1 tablespoon vegetable oil
½ cup chopped onion
2 cloves garlic, minced
¾ cup red lentils

Red lentils are soft and cook quickly; when properly cooked, they will be mostly whole, with some having dissolved to form a thick sauce. It's easy to overcook them and end up with a total puree (which will still taste fine), so check them often toward the end of the suggested cooking time.

For the rice: In a 1½-quart saucepan, bring 1½ cups of the water to a boil. Stir in the rice and return to a boil. Reduce the heat and simmer, covered, 20 minutes, or until the water is absorbed. Remove from the heat and stir in the cucumbers, scallion, coriander, margarine (if using), and ½ teaspoon of the salt.

For the lentils: While the rice is cooking, heat the oil over medium-high heat in a 1-quart saucepan. Add the onion and garlic, and cook, stirring, until softened. Add the remaining 1 cup water and bring to a boil. Add the lentils and return to a boil. Reduce the heat and simmer, covered, 10 minutes. Stir in the remaining ¼ teaspoon salt.

Serve the lentils over the rice.

SERVES 4

Calories:	313	Total fat:	6.8 g
Protein:	12.9 g	Saturated fat:	2.4 g
Carbohydrates:	50.8 g	Cholesterol:	0
Fiber:	5.4 g	Sodium:	430 mg

RATATOUILLE, BUTTER BEANS, AND BROWN RICE (V)

I make this dish using white eggplant, which does not turn the murky gray-brown that purple eggplant does when cooked. I cut all the vegetables into ¼-inch dice for a uniform consistency.

For the brown rice: In a 1-quart saucepan, bring 1⅔ cups of the water and ½ teaspoon of the salt to a boil. Add the rice and return to a boil. Reduce the heat and simmer, covered, 45 minutes.

For the ratatouille and butter beans: While the rice is cooking, heat the oil in a 3-quart saucepan over medium-high heat. Add the onion and garlic, and cook, stirring, until softened. Add the eggplant, zucchini, and bell pepper, and cook, stirring, until tender-crisp. Stir in the remaining ⅓ cup water and the tomato paste, parsley, marjoram, thyme, and rosemary. Cook, stirring, until the water has evaporated and the mixture has thickened. Stir in the butter beans and cook until heated through. Stir in the remaining ¼ teaspoon salt and the pepper.

Serve the ratatouille and butter beans over the brown rice.

SERVES 4 TO 6

RICE
- 2 cups water, divided
- ¾ teaspoon salt, divided
- ¾ cup short-grain brown rice

RATATOUILLE AND BUTTER BEANS
- 3 tablespoons olive oil
- ½ cup diced onion
- 2 cloves garlic, minced
- 1 cup diced eggplant
- 1 cup diced zucchini
- ½ cup diced green bell pepper
- 2 tablespoons tomato paste
- 2 tablespoons chopped parsley
- ¼ teaspoon dried marjoram or oregano
- ⅛ teaspoon dried thyme
- ⅛ teaspoon dried rosemary, crumbled
- 1½ cups cooked or canned (drained and rinsed) butter beans
- ⅛ teaspoon pepper

Calories:	311	*Total fat:*	11.3 g
Protein:	8.7 g	*Saturated fat:*	1.6 g
Carbohydrates:	45.4 g	*Cholesterol:*	0
Fiber:	3.9 g	*Sodium:*	415 mg

❧ CABBAGE, KASHA, AND BEANS (E)(V)

2 tablespoons vegetable oil
3 cups shredded cabbage
⅓ cup sliced scallion
1 package (10 ounces) frozen baby lima beans, un-
   thawed
1½ cups cooked kasha
   (roasted buckwheat)
¼ teaspoon salt
¼ teaspoon pepper
½ cup chopped walnuts

I serve this dish with a spoonful of unflavored yogurt on top. The nuts are a subtle element that adds the perfect finishing touch.

In a large skillet, heat the oil over medium-high heat. Add the cabbage and scallion, and cook, stirring, until softened. Add the beans, kasha, salt, and pepper. Cook, stirring, until the beans are heated through. Stir in the walnuts.

SERVES 4

Calories:	304	*Total fat:*	15.4 g
Protein:	10.4 g	*Saturated fat:*	2.0 g
Carbohydrates:	34.1 g	*Cholesterol:*	0
Fiber:	7.9 g	*Sodium:*	233 mg

❧ SENEGAL STEW WITH MILLET (V)

2 tablespoons vegetable oil
2 cups coarsely chopped
   cabbage
1 cup chopped onion
2 cloves garlic, minced
½ teaspoon ground red
   pepper
½ teaspoon curry powder
¼ teaspoon dried thyme
1 can (14½ ounces) whole
   peeled tomatoes, un-
   drained
1 cup vegetable (or
   chicken) broth or water

This unusual and delicious stew is slightly spicy, with a hint of curry. The peanut butter adds a wonderful consistency and a subtle flavor to the mixture. Be sure the cabbage has wilted before adding the other ingredients.

In a 3-quart saucepan, heat the oil over medium-high heat. Add the cabbage, onion, and garlic, and cook, stirring, until the cabbage is softened. Stir in the red pepper, curry powder, and thyme. Stir in the tomatoes, breaking them up with the back of a spoon. Add the broth and peanut butter, and stir until completely smooth. Add the potato, rutabaga, carrot, and chickpeas. Bring to a boil, reduce the heat, and simmer, uncovered, 35 minutes. Serve over the millet.

SERVES 4

- 2 tablespoons smooth peanut butter
- 2 cups cubed sweet potato
- 1½ cups cubed rutabaga (yellow turnip)
- 1 cup sliced carrot (½-inch-long pieces)
- 1 cup cooked or canned (drained and rinsed) chickpeas
- 2 cups cooked millet

Calories:	349	Total fat:	13.0 g
Protein:	11.2 g	Saturated fat:	2.0 g
Carbohydrates:	50.4 g	Cholesterol:	0
Fiber:	8.8 g	Sodium:	407 mg

❧ HERBED CHICKPEAS WITH AROMATIC RICE (V)

Aromatic rice is not a specific brand, but rather a type of rice that has a wonderful flowery flavor. I find Jasmine rice to be the most flavorful of all rices; Indian basmati is a close second. Uncle Ben's has recently come out with a rice called Aromatica, which also works well in this recipe. Another "new" aromatic rice is called Jasmati.

For the rice: In a 2-quart saucepan, bring the water to a boil over high heat. Stir in the rice and return to a boil. Reduce the heat and simmer, covered, 20 to 25 minutes, or until the liquid has been absorbed. Stir in the salt.

RICE
- 2 cups water
- 1 cup white aromatic rice
- ½ teaspoon salt

CHICKPEAS
- 1 tablespoon vegetable oil
- ⅔ cup chopped onion
- 2 cloves garlic, minced
- ½ cup vegetable broth
- 2 tablespoons chopped fresh basil
- 1 tablespoon chopped fresh rosemary
- 1 tablespoon chopped fresh thyme

2¼ cups cooked or canned (drained and rinsed) chickpeas

¼ cup chopped parsley

¼ teaspoon pepper

For the chickpeas: While the rice is cooking, in a medium skillet, heat the oil over medium-high heat. Add the onion and garlic, and cook, stirring, until softened. Stir in the broth, basil, rosemary, and thyme; bring to a boil. Add the chickpeas and simmer, covered, 10 minutes, or until the liquid has evaporated. Stir in the parsley and pepper.

Serve the chickpeas over the rice.

SERVES 4

Calories:	363	*Total fat:*	6.0 g
Protein:	11.9 g	*Saturated fat:*	0.6 g
Carbohydrates:	65.4 g	*Cholesterol:*	0
Fiber:	7.5 g	*Sodium:*	487 mg

CANNELLINI WITH FENNEL AND SAUTÉED ARUGULA (E)(Q)(V)

¼ cup pignoli (pine nuts)

3 tablespoons olive oil, divided

2 bunches arugula, rinsed and trimmed

3 cloves minced garlic, divided

2 cups sliced fennel or celery

½ teaspoon dried rosemary, crumbled

2 cups cooked or canned (drained and rinsed) cannellini

¼ teaspoon pepper

Fennel has a licorice flavor. If you are not fond of it you can use celery instead. Arugula, sometimes called rocket, is a peppery-tasting salad green. Watercress and spinach are acceptable substitutes if arugula is not available.

Place the pignoli in a large skillet and cook over medium heat, shaking the pan frequently, until the nuts are browned. Remove from the skillet and set aside.

Pour 1 tablespoon of the olive oil into the skillet and heat over medium-high heat. Add the arugula and 1 clove of the minced garlic, and sauté until the arugula is softened;

remove from the skillet to a serving platter. Add the remaining 2 tablespoons oil to the skillet, along with the fennel, rosemary, and remaining 2 cloves garlic; sauté until tender-crisp. Stir in the cannellini and pepper; cook until warmed through. Serve the cannellini over the arugula and top with the browned pignoli.

SERVES 4

Calories:	291	*Total fat:*	17.0 g
Protein:	14.2 g	*Saturated fat:*	2.4 g
Carbohydrates:	28.0 g	*Cholesterol:*	0
Fiber:	12.5 g	*Sodium:*	245 mg

≈ RATATOUILLE PIE (V)

To me, the test of a really successful recipe is when someone asks for seconds. With this dish, everyone asks for seconds! You can prepare instant polenta, or follow the directions for cooking polenta on pages 139–140.

Spread the polenta in the bottom of a greased 9-inch-square baking dish or 10-inch pie plate, forming a bottom crust. (If using a pie plate, spread the polenta only on the bottom of the plate, not up the sides.)

In a large skillet, heat the oil over medium-high heat. Add the onion and garlic, and cook, stirring, until the onion is softened. Remove 2 tablespoons of the onion from the pan and set aside.

Add the bell peppers to the onion

2 cups warm cooked polenta
3 tablespoons olive oil
1½ cups sliced onion
2 cloves garlic, minced
1 cup red bell pepper chunks
4 cups diced eggplant
2 cups chopped zucchini
1 can (16 ounces) whole tomatoes in thick puree
¼ cup chopped parsley
1 teaspoon distilled white vinegar
¾ teaspoon salt
½ teaspoon dried rosemary, crumbled
¼ teaspoon dried thyme
¼ teaspoon dried marjoram
⅛ teaspoon pepper

1½ cups cooked or canned (drained and rinsed) cannellini

and garlic in the skillet. Cook, stirring, until tender-crisp. Add the eggplant and zucchini, and cook, stirring, until slightly softened. Stir in the tomatoes with puree, breaking up the tomatoes with the back of a spoon. Stir in the parsley, vinegar, salt, rosemary, thyme, marjoram, and pepper. Spoon over the polenta crust.

Place the cannellini and reserved onion in the workbowl of a food processor fitted with a steel blade. Cover and process until smooth. Spoon the cannellini-onion mixture over the vegetables and gently spread with a spatula to form a top crust, leaving an occasional hole as a vent.

Bake in a 350° F oven for 40 minutes, or until the cannellini crust is lightly browned. Remove from the oven and let stand 10 minutes, to set.

SERVES 6

Calories:	217	*Total fat:*	7.7 g
Protein:	7.4 g	*Saturated fat:*	1.0 g
Carbohydrates:	32.0 g	*Cholesterol:*	0
Fiber:	8.5 g	*Sodium:*	358 mg

ꝏ SPAGHETTI SQUASH WITH VEGETABLE SAUCE (E)(V)

You can make this sauce with fresh cooked artichoke hearts, chopped coarsely, instead of the frozen ones that I use, but they certainly will be a lot more work. Don't substitute the marinated artichokes from a jar, since their flavor is too strong.

Cut the spaghetti squash in half widthwise and discard the seeds. Place the squash in a large saucepan and cover with water; bring to a boil over high heat. Reduce the heat and simmer, covered, 20 minutes, or until tender. Set aside.

Heat the oil in a medium skillet over high heat. Add the mushrooms and garlic; cook, stirring, until softened.

In a small bowl, stir together the water and cornstarch; stir in the broth. Add the broth mixture to the skillet along with the lemon juice, Madeira, and salt. Cook, stirring, until the mixture comes to a boil. Add the artichokes, olives, and parsley, and simmer 3 minutes.

Take the cooked squash and, using a fork, pull out the strands and discard them. Scoop out all the squash and place in a serving bowl. Spoon the sauce in the skillet over the squash. Sprinkle with the Parmesan (if using) and serve.

SERVES 4

1 spaghetti squash (3 pounds)
1 tablespoon olive oil
2 cups sliced mushrooms
3 cloves garlic, minced
1 tablespoon water
1 tablespoon cornstarch
1 cup vegetable (or chicken) broth
1 tablespoon fresh lemon juice
1 tablespoon Madeira
¼ teaspoon salt
1 package (9 ounces) frozen artichoke hearts, thawed and thoroughly drained
½ cup sliced pitted black olives
2 tablespoons chopped parsley
¼ cup grated Parmesan (optional)

Calories:	164	Total fat:	8.0 g
Protein:	5.9 g	Saturated fat:	2.1 g
Carbohydrates:	18.1 g	Cholesterol:	3.9 mg
Fiber:	5.1 g	Sodium:	410 mg

CAJUN LENTIL STEW (V)

1 tablespoon olive oil
¾ cup chopped onion
2 cloves garlic, minced
1½ cups water
1 bay leaf
½ cup lentils
1 can (15 ounces) whole peeled tomatoes, undrained
⅓ cup dry red wine
3 tablespoons tomato paste
1 teaspoon sugar
1 teaspoon dried basil
½ teaspoon dried oregano
¼ teaspoon dried thyme
¼ teaspoon salt
⅛ teaspoon pepper
2 cups chopped zucchini

This stew is a Louisiana-style chili. You can also serve it as a sauce over polenta or brown rice.

Heat the oil in a 3-quart saucepan over medium-high heat. Add the onion and garlic, and cook, stirring, until softened. Add the water and bay leaf, and bring to a boil. Stir in the lentils and return to a boil. Reduce the heat and simmer 30 minutes.

Stir in the tomatoes with their juice, breaking them up with the back of a spoon. Add the wine, tomato paste, sugar, basil, oregano, thyme, salt, and pepper. Bring to a boil; reduce the heat and simmer 10 minutes longer. Add the zucchini and simmer 15 minutes longer. Discard the bay leaf before serving.

SERVES 4 TO 6

Calories:	184	Total fat:	4.2 g
Protein:	8.6 g	Saturated fat:	0.6 g
Carbohydrates:	28.5 g	Cholesterol:	0
Fiber:	6.5 g	Sodium:	187 mg

✌ WHITE BEAN CHILI WITH WHEAT BERRIES (V)

For an even milder version of this fairly mild chili, use plain stewed tomatoes instead of the Mexican ones. Add ground red pepper to taste for a hotter chili.

In a 1-quart saucepan, bring 2 cups of the water to a boil over high heat. Add the beans and boil 2 minutes. Remove from the heat and let stand, covered, 30 minutes. Drain and rinse; set aside.

In a 4-quart saucepan, heat the oil over medium-high heat. Add the onion and garlic, and sauté until softened. Stir in the chili powder, paprika, cumin, and cinnamon. Add the remaining 3 cups water and bring to a boil. Stir in the drained beans and the wheat berries. Bring to a boil, reduce the heat, and simmer, covered, 1½ hours. Stir in the tomatoes and break them up with the back of a spoon. Add the tomato paste, bay leaf, salt, and oregano. Simmer 15 minutes longer. Discard the bay leaf before serving.

SERVES 4 TO 6

5 cups water, divided
¾ cup cow (field) beans, rinsed
2 tablespoons vegetable oil
2 cups chopped onion
2 cloves garlic, minced
3 tablespoons chili powder
1 teaspoon paprika
1 teaspoon ground cumin
½ teaspoon ground cinnamon
¾ cup wheat berries (whole-grain wheat)
1 can (14½ ounces) Mexican-style stewed tomatoes, undrained
3 tablespoons tomato paste
1 bay leaf
½ teaspoon salt
½ teaspoon dried oregano

Calories:	331	Total fat:	9.1 g
Protein:	13.4 g	Saturated fat:	1.5 g
Carbohydrates:	53.5 g	Cholesterol:	0
Fiber:	7.9 g	Sodium:	508 mg

❧ RED, RED CHILI (V)

1½ tablespoons vegetable oil
1½ cups chopped onion
1½ cups chopped red bell
  pepper
3 cloves garlic, minced
3 tablespoons chili powder
1 teaspoon paprika
½ teaspoon ground cinna-
  mon
½ teaspoon ground cumin
½ teaspoon dried oregano
3 cups chopped tomato
1 cup cranberry juice
1 tablespoon red wine vinegar
1¾ cups cooked or canned
  (drained and rinsed) red
  kidney beans
1¾ cups cooked or canned
  (drained and rinsed)
  black (turtle) beans
1 cup cooked bulgur or
  cracked wheat
1 teaspoon salt

Using red bell pepper instead of the usual green variety gives this chili a subtle flavor and adds to the intensely red color. The cranberry juice rounds off the flavors perfectly.

Heat the oil in a 4-quart saucepan over medium-high heat. Add the onion, bell pepper, and garlic. Cook, stirring, until softened. Remove from the heat and stir in the chili powder, paprika, cinnamon, cumin, and oregano until absorbed.

Add the tomato, cranberry juice, and vinegar. Stir until the tomato and spices are well combined. Return to the heat and bring to a boil. Reduce the heat and simmer 5 minutes. Stir in both types of beans and simmer, uncovered, 15 minutes. Stir in the bulgur and salt, and simmer 5 minutes longer.

SERVES 6

Calories:	292	*Total fat:*	4.7 g
Protein:	13.0 g	*Saturated fat:*	0.7 g
Carbohydrates:	51.6 g	*Cholesterol:*	0
Fiber:	14.3 g	*Sodium:*	498 mg

❧ QUICK VEGETABLE CHILI (E)(Q)(V)

This is a stew-type chili. If you like thicker chili, simmer it uncovered instead of covered. For a meaty texture, you can stir cooked bulgur or TVP (textured vegetable protein) into the chili. I like to serve it topped with a large dollop of unflavored yogurt.

In a 3-quart saucepan, heat the oil over medium-high heat. Stir in the onion, celery, carrot, bell pepper, and garlic. Cook, stirring, until the onion is softened. Stir in the chili powder, salt, cumin, red pepper flakes (if using), and oregano until absorbed. Add the crushed tomatoes and beans. Cover and simmer 20 minutes, or until the vegetables are tender.

SERVES 6

1½ tablespoons vegetable oil
1 cup chopped onion
1 cup chopped celery
1 cup diced carrot
½ cup chopped red or green bell pepper
3 cloves garlic, minced
3 tablespoons chili powder
¾ teaspoon salt
¾ teaspoon ground cumin
½ teaspoon red pepper flakes (optional, for a spicy chili)
½ teaspoon dried oregano
1 can (28 ounces) crushed tomatoes
2 cups cooked or canned (drained and rinsed) black (turtle) beans

Calories:	175	Total fat:	4.8 g
Protein:	7.4 g	Saturated fat:	0.7 g
Carbohydrates:	26.4 g	Cholesterol:	0
Fiber:	9.0 g	Sodium:	545 mg

❧ CURRIED SPINACH, TOFU, PEAS, AND POTATOES (V)

Use less than the ¼ teaspoon ground red pepper called for below if you like your curry mild (my tasters found this recipe a little too spicy). A more authentic version of

3 tablespoons vegetable oil
1 cup chopped onion
3 cloves garlic, minced
2 tablespoons curry powder

1 teaspoon ground cinnamon

1 teaspoon ground coriander

1 teaspoon ground cardamom

½ teaspoon ground turmeric

½ teaspoon ground cumin

¼ teaspoon ground red pepper

1½ cups water

2 cups peeled, diced new potato

1 cup fresh or frozen peas

1 bunch (10 ounces) fresh spinach, thoroughly rinsed, destemmed, and coarsely chopped

4 cups (24 ounces) diced tofu

½ teaspoon salt

the dish would use unsweetened coconut milk (the kind you can buy canned in stores that sell Middle Eastern or Oriental groceries) instead of the water, and ghee instead of the vegetable oil. Both of these substitutes, however, would add saturated fats (and, in the case of ghee, cholesterol) to the recipe.

Heat the oil in an 8-quart saucepan or stockpot over medium-high heat. Add the onion and garlic, and cook, stirring, until softened. Lower the heat and stir in the curry powder, cinnamon, coriander, cardamom, turmeric, cumin, and red pepper until absorbed.

Stir in the water and bring to a boil. Add the potato and peas, and simmer, uncovered, 30 minutes. Stir in the spinach, tofu, and salt; simmer 15 minutes longer.

SERVES 4 TO 6

Calories:	346	*Total fat:*	17.8 g
Protein:	19.5 g	*Saturated fat:*	2.5 g
Carbohydrates:	32.3 g	*Cholesterol:*	0
Fiber:	9.6 g	*Sodium:*	373 mg

❧ CHICKPEAS AND VEGETABLE MADRAS (V)

¾ cups dried chickpeas

3 tablespoons ghee (see page 79) or vegetable oil

2 cups coarsely chopped cabbage

1½ cups chopped onion

I like to serve this dish with Dal (see page 79) and Rice Pilau (page 121) to make a complete Indian meal. In India, Madras-style dishes are medium hot (vindaloo are the hottest). Depending on your own tolerance for spicy food, you may want to start with less ground red pepper than what is given below and only gradually increase the amount. For a shortcut, use 1½ cups drained and rinsed canned chickpeas instead of dried, and then omit the 2½ hours of cooking called for after the dried beans are added. If you use vegetable oil instead of ghee, the saturated fats will be reduced significantly, and there will be no cholesterol.

Soak the chickpeas overnight in 6 cups of water; drain and set aside.

In an 8-quart saucepan or stockpot, heat the ghee over medium-high heat. Add the cabbage, onion, ginger, and garlic. Cook, stirring, until softened. Stir in the curry powder, cumin, turmeric, red pepper, and cardamom until absorbed.

Stir in the chickpeas and water; bring to a boil. Reduce the heat and simmer, covered, 2½ hours, or until the beans are tender, adding 1 to 2 cups additional water if it starts to boil away. Stir in the cauliflower, carrot, and peas. Simmer, covered, 40 minutes. Stir in the milk, coriander, and salt; simmer, covered, 20 minutes longer.

SERVES 4

1 tablespoon minced fresh ginger
4 cloves garlic, minced
1½ tablespoons curry powder
½ teaspoon ground cumin
½ teaspoon ground turmeric
½ teaspoon ground red pepper
¼ teaspoon ground cardamom
1½ cups water
1½ cups cauliflower florets
1 cup sliced carrot
1 cup fresh or frozen peas
½ cup soy milk (or dairy milk)
2 tablespoons chopped fresh coriander (cilantro)
¾ teaspoon salt

Calories:	577	Total fat:	14.8 g
Protein:	12.3 g	Saturated fat:	7.9 g
Carbohydrates:	39.6 g	Cholesterol:	34.0 mg
Fiber:	6.3 g	Sodium:	548 mg

⮜ ALOO GOBI (E)(V)

1½ tablespoons vegetable oil
1 tablespoon chopped fresh ginger
½ teaspoon cumin seeds
1½ teaspoons *garam masala* or curry powder
1 teaspoon ground coriander
1 teaspoon ground turmeric
¼ teaspoon ground red pepper
6 cups cauliflower florets (1 large head)
2 cups peeled, cubed potato (½-inch pieces)
2 cups tomato wedges
2 tablespoons chopped fresh coriander (cilantro)
¾ teaspoon salt
1 tablespoon water
1 cup fresh or frozen peas

I usually think of curries (dishes featuring a blend of Indian spices, not necessarily including curry powder) as very garlicky dishes. This one is an exception — no garlic, and no onions! You probably will have to adjust the amount of liquid and the cooking time, since the ripeness of the tomatoes and the type of potato you use can alter cooking times significantly. If your tomatoes are not ripe, you may want to add more than the 1 tablespoon of water called for. Baking potatoes will take less time to soften than boiling potatoes. Start checking for doneness 15 minutes after the peas are added. The potatoes should be soft, and the finished dish should have a stewlike consistency.

Heat the vegetable oil in a 4-quart saucepan over medium-high heat. Add the ginger and cumin seed. Cook, stirring, 30 seconds. Add the *garam masala,* ground coriander, turmeric, and red pepper. Cook, stirring, 30 seconds, or until absorbed.

Add the cauliflower and potato, and cook, stirring, until they are coated with the spices. Add the tomato wedges, fresh coriander, and salt; toss to combine. Add the water. Cook over high heat 5 minutes, until steam forms. Reduce the heat to low-medium, cover, and cook 10 minutes, stirring occasionally. Uncover and simmer 10 minutes. Add the peas and cook, uncovered, until the potato is tender.

SERVES 4 TO 6

Calories:	207	Total fat:	5.8 g
Protein:	7.5 g	Saturated fat:	0.1 g
Carbohydrates:	34.9 g	Cholesterol:	0
Fiber:	8.7 g	Sodium:	468 mg

∾ CURRIED CHANA WITH RICE PILAU (V)

Chana are small dried chickpeas; they take considerably less time to cook than regular chickpeas. If you can't find chana, add 2 cups drained and rinsed canned chickpeas in place of the soaked chana and reduce the simmering time from 1 hour to 40 minutes. Ghee, clarified butter, is the traditional fat used in Indian cooking; if you are a vegan, you can substitute vegetable oil (this will also cut the cholesterol to 0.3 mg and the saturated fat to 1.9 g).

For the chana: In a 1-quart saucepan, bring 2 cups of the water to a boil. Add the chana and boil 2 minutes. Remove from the heat and set aside for 1 hour. Drain.

In a 4-quart saucepan, melt the ghee over medium heat. Add the onion, ginger, and garlic; cook, stirring, until softened. Stir in the curry

CHANA

4 cups water, divided

1 cup chana, rinsed and picked over

3 tablespoons ghee or vegetable oil

1 cup chopped onion

1 tablespoon chopped fresh ginger

3 cloves garlic, minced

2 tablespoons curry powder

1 teaspoon ground coriander

1 teaspoon ground cumin

1 teaspoon ground turmeric

¼ teaspoon ground cinnamon

¼ teaspoon ground red pepper

3 cups coarsely chopped cabbage

2 cups peeled, cubed new
red potato
1½ cups fresh or frozen cut
green beans
1½ cups chopped tomato
1 teaspoon salt

PILAU
1 tablespoon ghee or vegetable oil
½ cup chopped onion
½ teaspoon coriander seeds
¼ teaspoon cumin seeds
6 whole peppercorns,
cracked
4 whole allspice
1 cup white rice (preferably aromatic)
2 cups vegetable (or
chicken) broth
1 cup fresh or frozen peas
Salt to taste

powder, coriander, cumin, turmeric, cinnamon, and red pepper
until absorbed. Stir in the chana and
the remaining 2 cups water. Simmer, covered, 1 hour. Add the cabbage, potato, green beans, tomato,
and salt. Simmer, covered, 1¼
hours longer.

For the pilau: Thirty minutes before the curried chana are ready,
melt the ghee in a 2-quart saucepan.
Stir in the onion, coriander seeds,
cumin seeds, peppercorns, and allspice. Cook until the onion is softened. Stir in the rice and sauté 2
minutes. Stir in the broth and bring
to a boil. Reduce the heat and simmer, covered, 15 minutes. Stir in the
peas, cover, and simmer 5 minutes
longer. Add the salt.

Serve the curried chana over the
rice pilau.

SERVES 6

Calories:	436	*Total fat:*	10.0 g
Protein:	11.4 g	*Saturated fat:*	5.1 g
Carbohydrates:	64.2 g	*Cholesterol:*	21.0 mg
Fiber:	9.3 g	*Sodium:*	613 mg

❧ NOODLES WITH PEANUT SAUCE (E)(Q)(V)

¼ cup soy sauce
3 tablespoons smooth pea-
nut butter
1 tablespoon red wine vinegar

I use ultra-thin, very fast-cooking white Japanese noodles called to-moshiraga somen for this dish. If you cannot find them, you can use angel hair pasta or cappelletti in-

stead. *This recipe is not for sodium watchers. Even with light soy sauce there are over 600 mg sodium per serving.*

Put the soy sauce, peanut butter, vinegar, tahini, sherry, sugar, and sesame oil into a blender container. Cover and blend until smooth; set aside.

Cook the noodles in salted boiling water according to package directions. Drain and rinse under cold water until cool. Place in a large bowl; add the peanut sauce and scallions, and toss until thoroughly combined.

SERVES 4 TO 6

1 tablespoon tahini (sesame paste)
2 teaspoons dry sherry
1½ teaspoons sugar
1 teaspoon hot sesame oil or chili paste
12 ounces *tomoshiraga somen* or angel hair pasta
⅓ cup thinly sliced scallion

Calories:	439	Total fat:	10.5 g
Protein:	15.7 g	Saturated fat:	1.6 g
Carbohydrates:	71.3 g	Cholesterol:	0
Fiber:	3.5 g	Sodium:	1,050 mg

❧ BANGKOK NOODLES (E)(Q)(V)

I developed this recipe at the suggestion of Pat Baird, the author of Quick Harvest *(Prentice Hall, 1991), when we were tasting meat replacements. I was thrilled with the flavor and texture of savory baked tofu and was wondering how to include it in this book when Pat came up with the idea for this recipe. Savory baked tofu, which is pressed and marinated in soy sauce, is available in health food stores. If you can't find it, use firm*

12 ounces angel hair pasta
¼ cup soy sauce
2 tablespoons ketchup
1 tablespoon sugar
½ teaspoon red pepper flakes
1 tablespoon vegetable oil
2 cups bean sprouts
½ cup sliced scallion
3 cloves garlic, minced
1 cup diced savory baked tofu
⅓ cup chopped peanuts

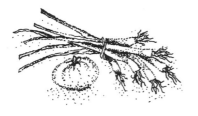

tofu instead. Those concerned with their sodium intake should probably steer clear of this dish. Using light soy sauce still results in over 600 mg sodium per serving.

Cook the pasta in salted water according to package directions; drain and set aside.

In a small bowl, stir together the soy sauce, ketchup, sugar, and red pepper flakes; set aside.

In a wok or large skillet, heat the oil over high heat. Add the bean sprouts, scallion, and garlic. Cook, stirring, until softened, about 1 minute. Add the noodles, tofu, and the soy sauce mixture. Cook, stirring, until heated through. Add the peanuts and toss until combined.

SERVES 4 TO 6

Calories:	482	*Total fat:*	12.1 g
Protein:	19.2 g	*Saturated fat:*	1.2 g
Carbohydrates:	75.6 g	*Cholesterol:*	0
Fiber:	4.9 g	*Sodium:*	1,192 mg

❧ FRAGRANT EGGPLANT WITH FRESH FIGS (E)(Q)(V)

¾ cup + 2 teaspoons water
2 tablespoons hoisin sauce
1½ tablespoons *mirin* or dry sherry
1 tablespoon honey
1 tablespoon ketchup

The exotic combination of flavors in this dish is a perfect blend of sweet and only slightly hot (you can add more pepper flakes if you like your eggplant spicy). I like to use Chinese eggplant, which are long and thin and lighter purple than the more commonly available type, but this recipe works well with either one.

In a small bowl, stir together ¾ cup of the water and the hoisin sauce, *mirin,* honey, ketchup, soy sauce, and red pepper flakes; set aside.

In a large skillet, heat 1 tablespoon of the oil over high heat. Add the bell pepper and cook, stirring, until tender-crisp; remove from the skillet and set aside. Add the remaining 1 tablespoon oil to the skillet; add the garlic and ginger, and stir-fry 30 seconds. Add the eggplant and stir-fry 1 minute. Stir in the hoisin mixture and cook, covered, 2 minutes.

In a small bowl, stir together the remaining 2 teaspoons water and the cornstarch. Stir into the eggplant mixture and stir-fry until the sauce has thickened. Add the reserved bell pepper and the figs, scallions, and walnuts. Cook, stirring, 1 minute longer, until the vegetables are heated through.

SERVES 4

1 tablespoon soy sauce
¼ teaspoon red pepper flakes
2 tablespoons vegetable oil, divided
2 cups cubed red or green bell pepper
4 cloves garlic, minced
1 tablespoon minced fresh ginger
6 cups diced eggplant
2 teaspoons cornstarch
4 fresh figs, cut into wedges
½ cup scallions, cut into 1½-inch pieces
½ cup walnuts

Calories:	264	Total fat:	13.2 g
Protein:	5.1 g	Saturated fat:	1.2 g
Carbohydrates:	34.8 g	Cholesterol:	0
Fiber:	7.6 g	Sodium:	423 mg

✺ BROCCOLI WITH GARLIC SAUCE (E)(Q)(V)

1 large bunch broccoli
¼ cup water
2 teaspoons cornstarch
3 tablespoons ketchup
1 tablespoon dark or black
   soy sauce
1 tablespoon sugar
2 teaspoons *mirin* or dry
   sherry
1 teaspoon rice vinegar or
   distilled white vinegar
3 tablespoons vegetable oil
⅓ cup sliced scallion
3 dried chili peppers,
   crumbled
2 teaspoons minced fresh
   ginger
3 cloves garlic, minced
2 cups (12 ounces) diced
   tofu

This yummy dish is moderately spicy. Adjust the number of chilies to your own taste. I like to serve this with brown rice instead of white.

Separate the broccoli florets from the main stalk. Peel the stalk to remove the tough outer skin. Slice the stalk into ¼-inch pieces. Rinse and drain the florets and sliced stalk; set aside.

In a small bowl, stir together the water and cornstarch until smooth. Stir in the ketchup, soy sauce, sugar, *mirin*, and vinegar; set aside.

In a wok or large skillet, heat the oil over high heat. Add the scallion, chili peppers, ginger, and garlic; cook, stirring, 30 seconds. Add the broccoli and cook, stirring, until the broccoli is tender-crisp.

Stir in the tofu and sauce, and cook, stirring, until the sauce has thickened and the tofu is heated through.

SERVES 4

Calories:	257	*Total fat:*	14.9 g
Protein:	17.6 g	*Saturated fat:*	1.8 g
Carbohydrates:	18.4 g	*Cholesterol:*	0
Fiber:	6.2 g	*Sodium:*	431 mg

≈ SPICY TOFU WITH CLOUD EARS (Q)(V)

Cloud ears are also called tree fungus or dried fungus. They can be found in most Oriental grocery stores. Rehydrate them by soaking in boiling water. Don't discard the soaking liquid; it's full of flavor.

In a small bowl, combine the boiling water with the cloud ears; let stand 5 minutes.

In a small bowl, stir together the 1 tablespoon water and the cornstarch. Stir in the hoisin sauce, *mirin,* soy sauce, and ketchup; set aside.

In a wok or large skillet, heat the oil. Add the ginger, garlic, and red pepper flakes. Cook, tossing, 30 seconds. Add the tofu, bean sprouts, and scallion, and stir-fry 1 minute longer. Add the bamboo shoots, reserved cloud ears with their soaking liquid, and cornstarch mixture. Cook, stirring, until thickened and heated through.

SERVES 4

¼ cup boiling water
3 tablespoons cloud ears
1 tablespoon water
2 teaspoons cornstarch
3 tablespoons hoisin sauce
2 tablespoons *mirin* or dry sherry
1 tablespoon soy sauce
1 tablespoon ketchup
2 tablespoons vegetable oil
2 teaspoons minced fresh ginger
2 cloves garlic, minced
½ teaspoon red pepper flakes
4 cups (24 ounces) diced tofu
½ cup mung bean or soybean sprouts
⅓ cup sliced scallion
1 can (8 ounces) sliced bamboo shoots, drained

Calories:	336	Total fat:	19.3 g
Protein:	23.5 g	Saturated fat:	2.8 g
Carbohydrates:	22.4 g	Cholesterol:	0
Fiber:	9.6 g	Sodium:	423 mg

❧ SESAME TOFU AND VEGETABLES (E)(Q)(V)

1½ tablespoons vegetable oil
1 tablespoon finely julienned fresh ginger
1 jalapeño pepper, seeded and cut into thin shreds
4 cloves garlic, minced
2 cups julienned carrot
2 cups julienned celery
1 cup julienned green bell pepper
2 packages (6 ounces each) savory baked tofu, thinly sliced
1½ tablespoons soy sauce
1 tablespoon dry sherry
1 teaspoon sesame oil
1 tablespoon sesame seeds

Savory baked tofu is sold in the refrigerator cases of natural food stores. Pressed and marinated in soy sauce, it's very different from regular tofu. Its consistency is not soft like that of regular tofu, but chewy — almost like Gouda or Swiss cheese. Serve this dish over cooked brown rice.

In a wok or large skillet, heat the oil over high heat. Add the ginger, jalapeño pepper, and garlic, and cook, stirring, 10 seconds. Add the carrot and cook, stirring, 30 seconds. Add the celery and bell pepper, and cook, stirring, until the vegetables are just tender-crisp. Add the tofu, soy sauce, and sherry, and cook, stirring, about 30 seconds. Add the sesame oil and seeds, and cook, stirring, until heated through.

SERVES 4

Calories:	177	*Total fat:*	11.3 g
Protein:	8.8 g	*Saturated fat:*	1.5 g
Carbohydrates:	12.0 g	*Cholesterol:*	0
Fiber:	4.6 g	*Sodium:*	466 mg

❧ CHOW MEIN (E)(Q)(V)

This is the type of Chinese food I remember from my youth — long before anyone had ever heard the word Szechuan. Serve it with rice and crispy noodles. The amounts of cornstarch and broth needed may vary, depending on how much liquid your vegetables give off as you stir-fry them. If the mixture is too thick, stir in a little more broth; if it's too thin, stir extra cornstarch into a little cold water and then add it to the vegetables, cooking until thickened.

⅓ cup vegetable (or chicken) broth
1 tablespoon soy sauce
1 tablespoon *mirin* or dry sherry
1 teaspoon sugar
1½ tablespoons cornstarch
2 tablespoons vegetable oil
2 cups sliced onion, cut lengthwise
3 cloves garlic, minced
4 cups sliced Chinese cabbage or bok choy, cut crosswise
3 cups mung bean sprouts
2 cups sliced celery, cut diagonally
¾ cup sliced scallion

In a small bowl, mix the broth, soy sauce, *mirin*, and sugar. Stir in the cornstarch until dissolved; set aside.

In a wok or large skillet, heat the oil over high heat. Add the onions and garlic, and cook, stirring, 30 seconds. Add the Chinese cabbage, bean sprouts, celery, and scallion, and cook, stirring, until tender-crisp. Add the cornstarch mixture and cook, stirring, until the sauce is thickened.

SERVES 4

Calories:	145	*Total fat:*	7.4 g
Protein:	5.5 g	*Saturated fat:*	0.4 g
Carbohydrates:	16.5 g	*Cholesterol:*	0
Fiber:	5.6 g	*Sodium:*	438 mg

↝ LO MEIN (E)(Q)(V)

12 ounces dried Chinese lo
  mein noodles or angel
  hair pasta
6 dried shiitake mush-
  rooms
⅔ cup boiling water
¼ cup vegetable (or
  chicken) broth
3 tablespoons dark or
  black soy sauce
1 tablespoon *mirin* or dry
  sherry
1 tablespoon black bean
  sauce
½ teaspoon sugar
2 teaspoons cornstarch
2 tablespoons vegetable oil
1 clove garlic, minced
3 cups sliced Chinese cabbage or bok choy, cut
  crosswise
1 cup julienned carrot
½ cup chopped onion
½ cup julienned snow peas
½ cup bean sprouts
½ cup sliced scallion
¼ cup frozen peas

Black bean sauce can usually be found in your supermarket's Oriental department. If your cooked and drained noodles have been waiting more than a few minutes while you were preparing the vegetables, rinse them under cold water immediately before adding to the wok to make them easier to toss. Sodium watchers beware — this dish has a high sodium count.

Cook the noodles in boiling water until tender. Drain and set aside.

In a small bowl, soak the mushrooms in the boiling water until softened, about 10 minutes. Discard the tough stems and slice the mushrooms; set aside, reserving the soaking liquid. Add the vegetable broth, soy sauce, *mirin*, black bean sauce, and sugar to the soaking liquid. Stir in the cornstarch; set aside.

In a wok or large skillet, heat the oil over high heat. Add the garlic and cook, stirring, 10 seconds. Add the Chinese cabbage, carrot, onion, snow peas, bean sprouts, scallion, and peas, and cook, stirring, until the vegetables are tender-crisp. Stir in the cornstarch mixture and cook, stirring, until thickened. Stir in the reserved noodles and cook, stirring, until the noodles are heated through.

SERVES 4 TO 6

Calories:	346	*Total fat:*	7.9 g
Protein:	10.9 g	*Saturated fat:*	1.1 g
Carbohydrates:	58.1 g	*Cholesterol:*	0
Fiber:	6.8 g	*Sodium:*	878 mg

❧ PENNE FROM HEAVEN (E)(Q)(V)

The fresh sauce in this recipe gets its heavenly flavor from all the vegetables in it. Be sure to remove the tough strings from the top and bottom of the sugar snaps.

Cook the pasta according to package directions; drain.

While the pasta is cooking, heat the oil over medium-high heat in a 3-quart saucepan. Add the onion and garlic, and cook, stirring, until softened. Add the tomato, water, salt, and sugar. Cook, stirring often, 10 minutes. Add the broccoli, zucchini, sugar snaps, carrot, and parsley. Cook, stirring, until the vegetables are tender-crisp. Stir in the Parmesan. Serve over the pasta.

SERVES 4 TO 6

12 ounces penne or ziti
2 tablespoons olive oil
1½ cups chopped onion
3 cloves garlic, minced
3 cups chopped ripe plum
   tomato
½ cup water
½ teaspoon salt
½ teaspoon sugar
2 cups broccoli florets
2 cups chopped zucchini
1 cup sugar snaps, tough
   strings removed, or snow
   peas
½ cup julienned carrot
2 tablespoons chopped
   parsley
¼ cup grated Parmesan
   (optional)

Calories:	481	*Total fat:*	10.0 g
Protein:	17.2 g	*Saturated fat:*	2.1 g
Carbohydrates:	81.9 g	*Cholesterol:*	3.9 mg
Fiber:	8.7 g	*Sodium:*	395 mg

~ PASTA WITH BROCCOLI AND ZUCCHINI IN GARLIC BROTH (E)(Q)(V)

12 ounces spaghetti or lin-
  guine
2 cups vegetable (or beef)
  broth
1 tablespoon minced garlic
4 cups broccoli florets
2 cups sliced zucchini
1 tablespoon cornstarch
1 tablespoon water

This is one of the easiest, fastest meals I can think of. You can use any vegetables you have on hand instead of only the broccoli and zucchini. For most recipes I find that using a garlic press is fine, but for this one I recommend mincing by hand.

Cook the pasta according to package directions; drain and place in a large serving bowl.

While the pasta is cooking, stir together the broth and garlic in a 2-quart saucepan. Cook over high heat until reduced to 1½ cups, about 6 minutes. Add the broccoli and zucchini, and cook until just tender, about 2 minutes. Remove the vegetables from the pan and place on top of the pasta.

In a small bowl, stir together the cornstarch and water. Stir into the garlic broth and bring to a boil. Pour over the vegetables and pasta.

SERVES 4

Calories:	372	*Total fat:*	1.9 g
Protein:	15.0 g	*Saturated fat:*	0
Carbohydrates:	74.0 g	*Cholesterol:*	0
Fiber:	7.3 g	*Sodium:*	454 mg

❧ ZITI WITH WHITE BEAN MARINARA SAUCE (E)(V)

I like these beans best when served with medium or small pasta. Try using orzo, bow ties, or orecchiette instead of the ziti for a nice change.

In a 4-quart saucepan, heat the oil over medium-high heat. Add the onion and garlic, and cook, stirring, until softened. Add the tomatoes with their liquid, breaking them up with the back of a spoon. Stir in the wine, parsley, tomato paste, sugar, basil, oregano, salt, thyme, and pepper. Bring to a boil. Reduce the heat and simmer, uncovered, 30 minutes. Stir in the beans and simmer 5 minutes longer, or until the sauce reaches the desired thickness.

Ten minutes before the sauce is ready, add the ziti to a large pot of salted boiling water. Cook 10 minutes, or until the ziti reaches the desired doneness. Drain. Serve topped with the sauce.

SERVES 4 TO 6

1 tablespoon olive oil
1 cup chopped onion
3 cloves garlic, minced
1 can (28 ounces) whole Italian tomatoes, undrained
½ cup dry red wine
¼ cup chopped parsley
2 tablespoons tomato paste
2 teaspoons sugar
1 teaspoon dried basil
½ teaspoon dried oregano
¼ teaspoon salt
¼ teaspoon dried thyme
¼ teaspoon pepper
1¾ cups cooked or canned (drained and rinsed) small white beans
12 ounces ziti or pasta of choice

Calories:	529	Total fat:	5.6 g
Protein:	19.7 g	Saturated fat:	0.7 g
Carbohydrates:	96.8 g	Cholesterol:	0
Fiber:	11.3 g	Sodium:	517 mg

❧ PASTA AND BEANS WITH PESTO (E)(Q)(V)

¼ cup pignoli (pine nuts)
¼ cup olive oil
½ cup packed fresh basil
4 cloves garlic, minced
¼ teaspoon salt
⅛ teaspoon pepper
1½ cups *tubetti* or other medium-sized pasta
2 cups cooked or canned (drained and rinsed) Roman (cranberry) beans
¼ cup grated Asiago or Parmesan (optional)

I sometimes use a small cheese-filled pasta, such as tortellini or ravioletti, instead of plain pasta, to make this dish even more substantial.

Place the pignoli in a dry skillet and cook, over medium heat, until browned; set aside.

Put the oil, basil, garlic, salt, and pepper in a blender container or the workbowl of a food processor fitted with a steel blade. Cover and process until the basil is finely chopped; set aside.

Cook the *tubetti* according to package directions until al dente. Drain and return to the pot. Add the beans and pesto, and cook, tossing, until the beans and sauce are warmed. Add the cheese (if using) and pignoli and toss.

SERVES 4

Calories:	455	*Total fat:*	23.0 g
Protein:	16.6 g	*Saturated fat:*	4.0 g
Carbohydrates:	49.1 g	*Cholesterol:*	3.9 mg
Fiber:	6.9 g	*Sodium:*	331 mg

~ MIDSUMMER'S PASTA (E)(Q)(V)

This sauce is best when made with vine-ripened tomatoes in their prime. It should cook only long enough to become slightly thick, to ensure that the flavor of the tomatoes and fresh basil is preserved. I'm usually too lazy to peel the tomatoes, but if you do, the sauce will be even better. If really delicious tomatoes are not available, skip this sauce and go directly to one that uses canned tomatoes.

1½ tablespoons olive oil
1 cup chopped onion
2 cloves garlic, minced
4 cups chopped tomato
⅓ cup chopped fresh basil
½ teaspoon sugar
½ teaspoon salt
  Pinch red pepper flakes
12 ounces pasta, cooked according to package directions

Heat the oil in a 2-quart saucepan over medium heat. Add the onion and garlic, and cook, stirring, until the onion is soft. Add the tomato, basil, sugar, salt, and red pepper flakes. Bring to a boil. Cook, stirring occasionally, 10 minutes, or until the desired thickness is reached. Serve over the pasta.

SERVES 4 TO 6

Calories:	380	Total fat:	2.5 g
Protein:	11.6 g	Saturated fat:	0.3 g
Carbohydrates:	70.0 g	Cholesterol:	0
Fiber:	3.9 g	Sodium:	267 mg

❧ BAKED ZITI AND EGGPLANT WITH BASIL-TOMATO SAUCE

2 tablespoons olive oil, divided

3 cups cubed eggplant

⅓ cup water

1 cup chopped onion

2 cloves garlic, minced

1 can (28 ounces) whole peeled Italian tomatoes

¼ cup tomato paste

1 teaspoon sugar

¼ teaspoon salt

⅛ teaspoon red pepper flakes

8 ounces ziti

½ cup chopped fresh basil

¼ cup grated Parmesan

1 cup shredded mozzarella

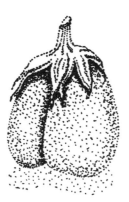

The tomato sauce for this dish is extremely fresh and delicious. If you cannot find fresh basil, you can substitute 2 teaspoons dried basil, but the sauce will lose its fresh taste. For a vegan version, serve over plain pasta and skip the Parmesan and the melted cheese step.

Preheat the oven to 400° F.

In a 3-quart saucepan, heat 1 tablespoon of the olive oil over medium-high heat. Add the eggplant and cook, stirring, 2 minutes, or until the eggplant starts to brown. Add the water and continue cooking until the eggplant is tender and the water is absorbed. Remove from the pan and set aside.

Add the remaining 1 tablespoon oil to the pan. Add the onion and garlic, and cook, stirring, until the onion is softened. Add the tomatoes, breaking them up with the back of a spoon. Add the tomato paste, sugar, salt, and red pepper flakes, and cook, stirring occasionally, 15 minutes.

While the sauce is cooking, cook the ziti in boiling salted water, according to package directions, until just barely al dente; drain. (The pasta will be cooking 10 minutes longer in the oven, so you want it just slightly underdone when you drain it.)

Stir the eggplant, basil, and Parmesan into the tomato sauce. Add the ziti and mix. Spoon into a 2-quart baking dish and sprinkle the top with the mozzarella. Bake 10 minutes, or until the cheese has melted and the pasta is heated through.

SERVES 4 TO 6

Calories:	459	*Total fat:*	15.6 g
Protein:	18.4 g	*Saturated fat:*	5.9 g
Carbohydrates:	62.9 g	*Cholesterol:*	26.2 mg
Fiber:	6.4 g	*Sodium:*	673 mg

⌦ VEGETABLE LASAGNE

A fair amount of work goes into this lasagne, but it's well worth every minute. This is a great dish to serve company. Even meat eaters are happy with a lasagne. Sodium watchers please note: all the cheese in this recipe contributes to a high sodium content.

Preheat the oven to 350° F.

For the sauce: In a 4-quart saucepan, heat the oil over medium-high heat. Add the onion and garlic, and cook, stirring, until softened. Add the eggplant and zucchini, and cook, stirring, until softened. Add the tomatoes and the puree, breaking up the tomatoes with the back of a spoon. Stir in the water, tomato paste, sugar, bay leaf, oregano,

SAUCE

1½ tablespoons olive oil
2 cups chopped onion
3 cloves garlic, minced
4 cups diced eggplant (½-inch pieces)
2 cups julienned zucchini
1 can (28 ounces) tomatoes in puree
1½ cups water
1 can (6 ounces) tomato paste
1 tablespoon sugar
1 bay leaf
1 teaspoon dried oregano
1 teaspoon dried basil
¼ teaspoon salt
¼ teaspoon dried thyme
¾ teaspoon pepper, divided

FILLING

   2 containers (1 pound
      each) drained cottage
      cheese
   1 package (8 ounces) moz-
      zarella, shredded, divided
   ½ cup chopped parsley,
      preferably flat leaf
   ¼ cup grated Parmesan
   12 lasagne noodles

basil, salt, thyme, and ¼ teaspoon of the pepper. Bring to a boil; reduce the heat and simmer 20 minutes. Discard the bay leaf.

For the filling: While the sauce is cooking, stir together in a large bowl the cottage cheese, all but ⅓ cup of the mozzarella, the parsley, the Parmesan, and the remaining ½ teaspoon pepper.

Cook the lasagne noodles according to package directions. Drain.

To assemble: Spread 1 cup of the tomato sauce in the bottom of a 9 × 13 × 2-inch baking dish. Place 6 cooked lasagne noodles on top of the sauce. Spread the cheese filling over the noodles. Spread half of the remaining sauce over the cheese. Top with the remaining 6 noodles. Spread the remaining sauce over the noodles and sprinkle with the reserved ⅓ cup mozzarella. Cover and bake 40 minutes. Uncover and bake 10 minutes longer. Let stand 10 minutes before serving.

SERVES 8

Calories:	435	*Total fat:*	10.9 g
Protein:	28.6 g	*Saturated fat:*	5.2 g
Carbohydrates:	54.5 g	*Cholesterol:*	27.5 mg
Fiber:	5.2 g	*Sodium:*	886 mg

POLENTA WITH EGGPLANT AND BLACK BEAN SAUCE (V)

Because this sauce is quick to cook, all the vegetables in it keep their fresh taste. I should warn you that the dish is fairly spicy, so you may want to cut back on the pepper flakes if you are faint of heart. Try serving this over pasta instead of polenta — it's terrific too. If you're in a hurry or don't want to do all the work for the polenta, you can use instant polenta, which takes only 5 minutes to prepare.

For the sauce: Heat the oil in a 3-quart saucepan over medium-high heat. Add the onion, bell pepper, and garlic, and cook, stirring, until softened. Add the eggplant and cook until a little browning occurs. Add the tomatoes and break them up with the back of a spoon. Stir in the water, tomato paste, parsley, basil, sugar, salt, and red pepper flakes. Cook over medium-high heat 10 minutes. Stir in the beans and cook 10 minutes longer.

For the polenta: In a 2-quart saucepan, bring the salt and 2½ cups of the water to a boil over high heat. While the water is boiling, stir together the remaining 1 cup water and the cornmeal. Add the cornmeal mixture to the boiling water all at once, stirring vigorously until completely combined. Reduce the heat to medium and simmer, stirring constantly (no cheating!), until the polenta pulls away from the

SAUCE

- 2 tablespoons olive oil
- ¾ cup chopped onion
- ½ cup chopped green bell pepper
- 2 cloves garlic, chopped
- 2 cups diced eggplant
- 1 can (15 ounces) whole peeled tomatoes, undrained
- ½ cup water
- ¼ cup tomato paste
- ¼ cup chopped parsley
- 1 teaspoon dried basil
- 1 teaspoon sugar
- ¼ teaspoon salt
- ½ teaspoon red pepper flakes
- 1½ cups cooked or canned (drained and rinsed) black (turtle) beans

POLENTA

- 3½ cups water, divided
- ¼ teaspoon salt
- 1¼ cups yellow cornmeal

sides of the pan as it is stirred, about 30 minutes.

Serve the polenta with the sauce on the side.

SERVES 4 TO 6

Calories:	375		Total fat:	8.2 g
Protein:	11.8 g		Saturated fat:	1.2 g
Carbohydrates:	65.7 g		Cholesterol:	0
Fiber:	13.1 g		Sodium:	490 mg

✍ POLENTA CHEESE PIE

POLENTA
2½ cups water
½ teaspoon salt
⅔ cup instant polenta

SAUCE
1 tablespoon olive oil
⅔ cup chopped onion
2 medium tomatoes, peeled, seeded, and chopped (about 2 cups chopped)
1 tablespoon tomato paste

FILLING
1 cup ricotta
¼ cup grated Parmesan
¼ cup chopped Italian (flat-leaf) parsley
¼ teaspoon pepper
1 teaspoon minced garlic (done by hand, not in a garlic press)

This is a very delicate and delicious dish, ideal for lunch or a light supper. I use instant polenta, available in gourmet stores, for this recipe. If you can't find it, you can make polenta from cornmeal by following the directions in the recipe for Porcini Polenta, page 158. To peel a tomato, place in boiling water for 1 or 2 minutes and then plunge into ice water or run under cold water. The skin should come off easily. If you have a gas stove, you can put the tomato on a fork and hold it over a high flame until the skin is blistery all over; then peel.

Preheat the oven to 400° F. Grease a 9-inch springform pan.

For the polenta: Bring the water and salt to a boil in a 1½-quart pot. Gradually stir in the instant polenta. Cook, stirring constantly, over medium heat 10 minutes, or until very thick.

For the sauce: In a small skillet, heat the oil. Add the onion and sauté until softened. Add the tomatoes and cook, stirring occasionally, 5 minutes, or until almost all the liquid has evaporated. Stir in the tomato paste and set aside.

For the filling: In a medium bowl, mix the ricotta, Parmesan, parsley, and pepper; set aside.

To assemble: Spread the polenta over the bottom of the prepared pan and 1 inch up the sides. Sprinkle with the garlic. Spread the ricotta mixture over the polenta. Spread the sauce over it. Top with the Gouda and mozzarella.

Bake 25 minutes, or until the cheeses are melted and browned. Let stand 5 minutes before removing springform sides.

SERVES 6

TOPPINGS

½ cup shredded Gouda

½ cup shredded mozzarella

Calories:	240	*Total fat:*	13.1 g
Protein:	12.2 g	*Saturated fat:*	7.1 g
Carbohydrates:	18.6 g	*Cholesterol:*	40.3 mg
Fiber:	2.6 g	*Sodium:*	403 mg

✑ TOSTADAS (E)(V)

Tostadas are basically open-faced tacos. You can also use this bean filling for enchiladas or tacos. I use marinated jalapeño peppers (the kind that come in a jar), and they are truly fiery, even after the seeds have been discarded. Use them at your discretion.

2 teaspoons vegetable oil

¼ cup minced onion

3 cloves garlic, minced

1¾ cups cooked or canned (rinsed and drained) pinto or Roman (cranberry) beans

⅓ cup water

1 tablespoon minced jala-
  peño pepper (optional)
½ teaspoon ground cumin
4 tostada shells
½ cup shredded cheddar
  (optional)
1 cup shredded lettuce
½ cup diced tomato
¼ cup chopped onion or
  scallion
  Tomato salsa, fresh or
  jarred

Heat the oil in a 1-quart saucepan over medium-high heat. Add the onion and garlic, and cook, stirring, until the onion is softened. Remove from the heat. Add the beans and mash with a fork until half the beans are pureed and the rest are slightly broken up. Stir in the water, jalapeño pepper (if using), and cumin. Return to the heat and cook, stirring, about 4 minutes, or until the liquid has been absorbed and the mixture is heated through.

Spread a quarter of the bean mixture over each tostada shell. Top each with 2 tablespoons cheddar (if using), ¼ cup lettuce, 2 tablespoons diced tomato, and 1 tablespoon onion. Serve with the tomato salsa.

SERVES 4

Calories:	297	*Total fat:*	15.4 g
Protein:	12.0 g	*Saturated fat:*	4.5 g
Carbohydrates:	31.3 g	*Cholesterol:*	14.9 mg
Fiber:	8.8 g	*Sodium:*	283 mg

❧ MEXICAN PIZZA

CRUST
½ cup very warm water
  (105–115° F)
½ teaspoon sugar
1 package active dry yeast
2¼ to 2¾ cups all-purpose
  flour
⅓ cup white or yellow
  cornmeal
1½ teaspoons salt

I make four individual pizzas, because it's easier to get small pizzas into and out of the oven. I bake the first pizza and share it with my guests while I have the next one baking. If you are brave or skilled, you can use this recipe to make one large pizza. For variety, add toppings such as sliced olives, onions, scallions, and jalapeño peppers.

For the crust: In a glass measuring cup, stir together the warm water and sugar. Stir in the yeast and let stand until ¼ inch of bubbly foam forms on top. (This process is called proofing the yeast, and if the foam doesn't appear, it means that the yeast has not been activated. Discard this batch and try again, double-checking the date on the yeast package and making sure the water is not overly warm or cool.)

In a large bowl, stir together 1½ cups of the all-purpose flour and the cornmeal and salt. Stir in the yeast mixture, water, and oil. Stir in ¼ cup more of the flour to make a soft dough. Turn out onto a well-floured board and knead in as much of the remaining flour as needed to form a dough that is smooth and elastic. Knead at least 10 minutes; it's important that this dough be well kneaded.

Place the dough in a greased bowl, turning so that all sides have been greased, and cover with greased plastic wrap. Put in a warm, draft-free spot to rise until doubled in bulk (about 1 hour).

For the filling: Heat the oil in a 1-quart saucepan. Add the onion and cook, stirring, until softened. Add the beans, water, cumin, and salt, and cook over medium-high heat until the water has evaporated (about 7 minutes). Place in the workbowl of a food processor fitted with a steel blade. Add the garlic. Cover and process until smooth; set aside.

For a gas oven: Remove the racks

½ cup water
1 tablespoon olive oil

FILLING
1 tablespoon vegetable oil
⅓ cup chopped onion
1¼ cups cooked pinto beans
¼ cup water
¼ teaspoon ground cumin
¼ teaspoon salt
2 cloves garlic, minced

TOPPINGS
½ to ¾ cup salsa (mild or hot, to taste) or taco sauce
½ cup shredded cheddar
½ cup shredded mozzarella

from the oven and place a large baking sheet on the oven floor (with the rim toward the back of the oven).

For an electric oven: Move a rack onto the lowest rung and place a large baking sheet on it (with the rim toward the back of the oven).

Preheat the oven (and baking sheet) to 450° F.

Punch down the dough; divide into fourths. On a surface lightly covered with flour, pat 1 piece of the dough as flat as possible. Make a fist and place the flattened dough on it. Gently lift and spread your fingers to stretch the dough. When you have stretched it a little, start using both hands to continue the stretching. When the center of the dough is thin, hold the dough with both hands and rotate the dough, squeezing the edges, making them thinner and the circle larger (you can do this while holding the dough suspended in the air). Continue these motions until you have an 8-inch circle.

Place on a baking sheet well dusted with cornmeal. Spread with a quarter of the bean puree (this should be a fairly thin layer). Top with 2 to 3 tablespoons of the salsa. Sprinkle with 2 tablespoons shredded cheddar and 2 tablespoons shredded mozzarella (and with any additional toppings you choose).

Open the oven door and transfer the pizza to the preheated baking sheet inside the oven (you may have to use a jerking motion to persuade the pizza to slide onto the sheet). Bake 12 minutes, or until the crust is crisp. (The edges will not be as brown as in a regular pizza, but the bottom should be browned, the edges crisp, and the cheese melted.)

To remove from the oven, slide the baking sheet on which you assembled the pizza under the baked pizza and, with a long spatula, lift the pizza onto it, leaving the preheated baking sheet in the oven for the next pizza.

Repeat the same process for the next 3 pizzas.

SERVES 4

Calories:	585	Total fat:	16.0 g
Protein:	21.4 g	Saturated fat:	5.7 g
Carbohydrates:	87.6 g	Cholesterol:	23.3 mg
Fiber:	9.4 g	Sodium:	389 mg

Side Dishes

❧ ZUCCHINI-APPLE BULGUR (E)(Q)(V)

¾ cup water
⅔ cup apple juice
¾ cup bulgur or cracked
   wheat
3 tablespoons pignoli (pine
   nuts)
1½ tablespoons vegetable oil
1 cup chopped apple
   (peeled or unpeeled, to
   taste)
1 cup chopped zucchini
½ teaspoon salt
¼ teaspoon ground carda-
   mom
¼ teaspoon ground cinna-
   mon

If you don't have cardamom on hand, you can use cinnamon or a pinch of nutmeg instead. This dish is slightly sweet, with a nice contrast of textures.

In a 2-quart saucepan, over high heat, bring the water and apple juice to a boil. Add the bulgur and return to a boil. Reduce the heat and simmer 20 minutes, or until the liquid is absorbed.

While the bulgur is cooking, lightly toast the pignoli in a large dry skillet over medium heat. Remove from the skillet and set aside.

Add the oil to the skillet and heat over medium-high heat. Add the apple and zucchini, and cook, stirring, until softened. Stir in the salt, cardamom, and cinnamon. Add the cooked bulgur and reserved pignoli.

SERVES 4

Calories:	191	Total fat:	4.3 g
Protein:	5.7 g	Saturated fat:	0.8 g
Carbohydrates:	35.7 g	Cholesterol:	0
Fiber:	7.0 g	Sodium:	403 mg

❧ COUSCOUS WITH THREE PEPPERS (E)(Q)(V)

If you have only one kind of pepper on hand, you can still make this very quick side dish, but of course it will not be as spectacular to look at. Omit the jalapeño pepper if you like your foods mild.

Heat the oil in a 2-quart saucepan over medium-high heat. Add all the bell pepper and the onion and jalapeño pepper. Cook, stirring, until softened. Add the broth and bring to a boil. Add the couscous and return to a boil. Remove from the heat and let stand, covered, 5 minutes. Stir in the tomato and parsley.

SERVES 6 TO 8

2 tablespoons olive oil
½ cup chopped red, orange, or yellow bell pepper
½ cup chopped green bell pepper
½ cup chopped onion
2 teaspoons minced jalapeño pepper
1¾ cups vegetable (or chicken) broth
1 cup couscous
½ cup chopped tomato
2 tablespoons chopped parsley

Calories:	158	Total fat:	5.4 g
Protein:	4.0 g	Saturated fat:	0.9 g
Carbohydrates:	24.5 g	Cholesterol:	0
Fiber:	5.9 g	Sodium:	459 mg

❧ WHEAT BERRIES WITH CELERY AND GREEN BEANS (E)(Q)(V)

This very-easy-to-make dish is extremely flavorful. No salt is called for, because I am assuming that you have salted the wheat berries after you cooked them. You can use frozen green beans, if the fresh are out of season.

1½ tablespoons vegetable oil
3 tablespoons minced shallot
2 cups fresh green beans, cut into 1½-inch pieces
1 cup sliced celery

2½ cups cooked wheat ber-
ries (whole-grain wheat)
1 teaspoon sesame oil

Over medium-high heat, heat the oil in a 2-quart saucepan. Add the shallot and cook, stirring, until softened. Add the green beans and celery, and cook, stirring, until tender-crisp. Stir in the whole-grain wheat and cook, stirring, until heated through. Stir in the sesame oil.

SERVES 8

Calories:	114	Total fat:	6.2 g
Protein:	2.5 g	Saturated fat:	0.7 g
Carbohydrates:	13.4 g	Cholesterol:	0
Fiber:	3.3 g	Sodium:	330 mg

❧ WHEAT BERRIES WITH OYSTER MUSHROOMS (E)

1½ tablespoons olive oil
1 tablespoon minced shal-
lot
1 cup oyster mushrooms,
chopped
1½ cups cooked wheat ber-
ries (whole-grain wheat)
2 tablespoons chopped
parsley
1 tablespoon grated Parmesan
¼ teaspoon salt
⅛ teaspoon pepper

Oyster mushrooms are ear-shaped and grayish-brown. They have a delicate flavor. If you can't find them fresh, Frieda's Finest (see Sources, pages 63–64) sells them dried. Rehydrate ⅓ cup of the dried mushrooms according to package directions and then proceed with the recipe as written (saving the soaking liquid for soup or rice).

Heat the oil in a large skillet over medium-high heat. Add the shallot and sauté for 10 seconds; add the mushrooms and cook, stirring, until tender. Add the wheat berries and cook, stirring, until heated through. Add the parsley, Parmesan, salt, and pepper.

SERVES 4

Calories:	150	Total fat:	5.9 g
Protein:	4.6 g	Saturated fat:	1.0 g
Carbohydrates:	18.0 g	Cholesterol:	0.9 mg
Fiber:	4.4 g	Sodium:	426 mg

❧ WHEAT BERRIES AND ORIENTAL VEGETABLES (E)(V)

If you are in a hurry and have forgotten to cook your wheat berries ahead of time, you can stir in cooked quinoa (which takes only 20 minutes to prepare), or you can simply cook the vegetables and serve them over rice. Either alternative will be quicker than wheat berries, which take almost 2 hours to cook.

In a wok or large skillet, heat the oil over high heat. Add the garlic and cook, stirring, 10 seconds. Add the cabbage, bean sprouts, bell pepper, snow peas, and scallions. Cook, stirring, until tender-crisp. Stir in the soy sauce, sherry, and red pepper flakes (if using). Add the wheat berries and cook, stirring, until heated through.

SERVES 4 TO 6

2 tablespoons vegetable oil
2 cloves garlic, minced
2 cups shredded Chinese cabbage or bok choy
1 cup mung bean sprouts
½ cup julienned red bell pepper
½ cup chopped snow peas
⅓ cup scallions, cut into 1-inch pieces
1 tablespoon soy sauce
1 tablespoon dry sherry
¼ teaspoon red pepper flakes (optional)
1½ cups cooked wheat berries (whole-grain wheat)

Calories:	187	Total fat:	7.3 g
Protein:	6.1 g	Saturated fat:	1.1 g
Carbohydrates:	22.3 g	Cholesterol:	0
Fiber:	6.2 g	Sodium:	529 mg

❧ BROWN RICE WITH CURRIED FRUIT (V)

1½ tablespoons vegetable oil
½ cup chopped onion
1 tablespoon curry powder
1 teaspoon ground ginger
⅛ teaspoon ground cloves
⅛ teaspoon ground red
  pepper
½ cup apple juice
1 cup chopped mango
½ cup chopped peach
½ cup chopped apple
⅓ cup raisins
1 teaspoon cider vinegar
½ teaspoon salt
3 cups cooked brown rice
⅓ cup chopped walnuts

This mouth-tingling rice dish is a real treat. If you don't have fresh peaches, just increase the amount of chopped mango that you use, or vice versa.

In a 2-quart saucepan, heat the oil over medium-high heat. Add the onion and cook, stirring, until softened. Remove from the heat and stir in the curry powder, ginger, cloves, and red pepper until absorbed. Stir in the apple juice. Add the mango, peach, apple, raisins, vinegar, and salt. Cook uncovered over medium heat, stirring occasionally, until the fruits are cooked and the mixture is thickened, about 5 to 7 minutes. Stir in the rice and walnuts. Cook until heated through.

SERVES 6 TO 8

Calories:	301	*Total fat:*	8.6 g
Protein:	4.5 g	*Saturated fat:*	1.0 g
Carbohydrates:	54.5 g	*Cholesterol:*	0
Fiber:	4.9 g	*Sodium:*	438 mg

❧ RICE WITH EAST INDIAN FLAVORS (E)(Q)(V)

1 tablespoon oil
⅓ cup chopped onion
1½ tablespoons minced jalapeño pepper
1 tablespoon minced fresh
  ginger
2 cloves garlic, minced

I like Uncle Ben's Aromatica rice for this recipe because it's so flavorful and delicate. If you can't find it in the supermarket, try basmati (preferably Indian) or Jasmati rice. If all else fails, use regular white rice.

In a 1½-quart saucepan, heat the oil over medium-high heat. Add the onion, jalapeño pepper, ginger, garlic, and coriander seeds; cook, stirring, until softened. Stir in the water and rice; bring to a boil. Reduce the heat and simmer, covered, 15 to 20 minutes, or until the liquid is absorbed. Add the fresh coriander, mint, and salt; toss to combine.

SERVES 4

2 teaspoons coriander seeds, crushed
1½ cups water
¾ cup white rice
2 tablespoons chopped fresh coriander (cilantro)
1 tablespoon chopped fresh mint
½ teaspoon salt

Calories:	165	*Total fat:*	3.6 g
Protein:	2.8 g	*Saturated fat:*	0.5 g
Carbohydrates:	30.0 g	*Cholesterol:*	0
Fiber:	0.9 g	*Sodium:*	301 mg

❧ RICE PILAF (E)(Q)(V)

I use converted rice in this recipe because the grains stay separate better than do those of most other brands. This fairly standard pilaf goes well with any flavorful dish, and neither detracts from the other dish nor loses its own taste. The recipe calls for cultivated mushrooms, but you can jazz it up with more exotic and flavorful ones.

Heat the oil in a 2-quart saucepan over medium-high heat. Add the mushrooms and leek, and cook, stirring, until softened. Add the rice and cook, stirring, until some of the rice starts to brown. Add the broth and bring to a boil. Reduce the heat and simmer, covered, 20 minutes. Stir in the zucchini, tomato, and

1 tablespoon olive oil
1 cup sliced mushrooms
⅔ cup sliced leek (white and light green parts only), thoroughly rinsed
1 cup converted white rice
2⅓ cups vegetable (or chicken) broth
½ cup chopped zucchini
½ cup chopped tomato
¼ cup frozen peas
¼ cup chopped parsley
⅛ teaspoon pepper

peas. Cover and simmer 5 minutes longer. Stir in the parsley and pepper.

SERVES 6 TO 8

Calories:	160	*Total fat:*	3.0 g
Protein:	3.7 g	*Saturated fat:*	0.5 g
Carbohydrates:	29.7 g	*Cholesterol:*	0
Fiber:	1.6 g	*Sodium:*	320 mg

❧ RISOTTO WITH FRESH HERBS   (V)

2 cups vegetable (or chicken) broth
1¾ cups water
1½ tablespoons olive oil
¼ cup finely chopped shallot
1 cup arborio rice
¼ cup chopped fresh basil
¼ cup chopped parsley
2 teaspoons chopped fresh oregano
1 teaspoon chopped fresh thyme
⅓ cup grated Asiago or Parmesan (optional)

If you don't have the fresh herbs called for below, use 1 teaspoon dried basil, 1 tablespoon dried parsley flakes, ½ teaspoon dried oregano, and ¼ teaspoon dried thyme. Asiago is a very flavorful Italian cheese; it comes in both a soft and hard variety. The latter, which is good for grating, is the kind I use in this recipe.

In a 3-quart saucepan, heat the broth and water until simmering. Keep on low heat so that the liquids stay warm throughout the cooking time.

In a 2-quart saucepan, heat the oil over medium-high heat. Add the shallot and cook, stirring, until softened. Add the rice and stir to coat with the oil.

Gradually add the broth-water mixture to the rice and shallot, ¼ cup at a time, stirring constantly, until the rice has absorbed the liquid. You should make the next addition of liquid when you can draw a clear path on the bottom of the pan as you scrape through the rice with a wooden spoon. This will happen rather quickly at first and will take longer as you near the end of the cooking time. About three-quarters of the way through cooking, stir in the fresh herbs. Continue adding the broth ¼ cup at a time until the risotto is creamy, with al dente grains of rice. (You may not need all the liquid.) Stir in the cheese (if using).

SERVES 6

Calories:	175	*Total fat:*	5.2 g
Protein:	4.7 g	*Saturated fat:*	1.4 g
Carbohydrates:	26.8 g	*Cholesterol:*	3.8 mg
Fiber:	0.8 g	*Sodium:*	373 mg

ZUCCHINI AND YELLOW PEPPER RISOTTO (V)

This recipe can be changed in many easy ways and will still be a festive dish for company: use red, green, or orange peppers instead of the yellow; substitute chopped shallot or onion for the chives; or add yellow squash, peas, snow peas, or tomato instead of the zucchini.

In a 3-quart saucepan, heat the broth, water, and wine until simmering. Keep on low heat so that

1¾ cups vegetable (or chicken) broth
1½ cups water
½ cup dry white wine
2 tablespoons olive oil, divided
1 cup chopped zucchini
¾ cup chopped yellow bell pepper
2 tablespoons snipped chives
¼ teaspoon dried thyme

1 cup arborio rice
¼ cup grated Parmesan
(optional)

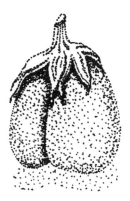

the liquid stays warm throughout the cooking time.

While the broth is heating, heat 1 tablespoon of the oil in a medium skillet over medium-high heat. Add the zucchini and bell pepper. Cook, stirring, until softened. Add the chives and thyme; remove from the heat and set aside.

In a 2-quart saucepan, over medium heat, heat the remaining 1 tablespoon oil. Add the rice and cook, stirring, until it is coated with the oil.

Gradually add the broth to the rice, ¼ cup at a time, stirring constantly, until the rice has absorbed the liquid. You should make the next addition of liquid when you can draw a clear path on the bottom of the pan as you scrape through the rice with a wooden spoon. This will happen rather quickly at first and will take longer as you near the end of the cooking time. Continue adding the broth ¼ cup at a time until the risotto is creamy, with al dente grains of rice. (You may not need all the liquid.) Stir in the sautéed vegetables and the Parmesan (if using).

SERVES 6

Calories:	193	*Total fat:*	6.0 g
Protein:	4.3 g	*Saturated fat:*	1.4 g
Carbohydrates:	26.8 g	*Cholesterol:*	2.1 mg
Fiber:	0.8 g	*Sodium:*	319 mg

ꝃ FRIED RICE (E)(V)

Try this recipe for fried rice with a much more interesting flavor than the type you usually get in a Chinese restaurant. It's also an excellent way to use up any leftover cooked rice. You can use extra tofu or add chopped scrambled egg to increase the protein value. If you are unable to find cloud ears, substitute ½ cup sliced mushrooms and sauté them with the cabbage.

⅓ cup boiling water
2 tablespoons cloud ears (tree fungus)
2 tablespoons dark or black soy sauce
1 tablespoon dry sherry
1 tablespoon hoisin sauce
3 tablespoons vegetable oil
1 cup shredded Chinese cabbage or bok choy
1 cup mung bean sprouts
⅔ cup chopped onion
2 teaspoons minced fresh ginger
2 cups cooked rice (if chilled, fluff with a fork before measuring)
½ cup frozen peas
½ cup (3 ounces) diced tofu (⅛-inch pieces)
⅓ cup chopped baby corn
¼ cup sliced scallion

Place the boiling water in a small bowl. Add the cloud ears and let stand 5 minutes. Drain and chop; set aside.

In a separate small bowl, stir together the soy sauce, sherry, and hoisin sauce until completely combined.

In a wok or large skillet, heat the oil over high heat. Add the Chinese cabbage, sprouts, onion, and ginger. Cook, stirring, until softened. Stir in the rice, peas, tofu, corn, scallion, and reserved cloud ears. Cook, stirring, until heated through. Add the soy sauce mixture and stir until absorbed by the fried rice.

SERVES 6 TO 8

Calories:	193	*Total fat:*	8.0 g
Protein:	5.3 g	*Saturated fat:*	1.1 g
Carbohydrates:	25.4 g	*Cholesterol:*	0
Fiber:	2.7 g	*Sodium:*	478 mg

❧ WHITE AND WILD RICE PILAF   (E)(V)

1 tablespoon vegetable oil
2 tablespoons minced shallot
½ cup wild rice, rinsed
1¾ cups vegetable broth or
   water
⅓ cup converted white rice
½ cup fresh or frozen peas
2 tablespoons chopped
   parsley
2 tablespoons finely
   chopped cranberries *or*
½ teaspoon grated lemon
   rind
⅛ teaspoon pepper

Wild rice cooks best when the water is unsalted. If you are using bouillon cubes to make the vegetable broth, cook the wild rice in plain water and add the cubes when you stir in the white rice. If you are using homemade broth, use unsalted broth, then salt to taste later.

In a 2-quart saucepan, heat the oil over medium-high heat. Add the shallot and cook, stirring, until softened. Stir in the wild rice and cook, stirring, 1 minute. Stir in the vegetable broth or water. Bring to a boil, reduce the heat, and simmer, covered, 40 minutes.

Add the white rice (and the bouillon cubes, if using) and simmer, covered, 10 minutes longer. Stir in the peas, parsley, cranberries, and pepper. Simmer, covered, 10 minutes longer, or until the liquid is absorbed.

SERVES 4 TO 6

Calories:	184	*Total fat:*	4.1 g
Protein:	5.5 g	*Saturated fat:*	0.6 g
Carbohydrates:	31.2 g	*Cholesterol:*	0.4 mg
Fiber:	3.2 g	*Sodium:*	430 mg

✒ ZUCCHINI, LEEKS, AND WILD RICE (E) (V)

This is a wonderful, elegant dish. If you can't find jicama, you can substitute Jerusalem artichokes, kohlrabi, or even pears.

In a large skillet, heat the oil over medium-high heat. Add the leek and cook, stirring, until softened. Add the zucchini and jicama and cook, stirring, until tender-crisp.

Stir in the wild rice, orange rind, salt, and pepper. Cook, stirring, until heated through.

SERVES 4

1½ tablespoons vegetable oil
½ cup sliced leek (white and light green parts only), thoroughly rinsed
¾ cup chopped zucchini
½ cup chopped jicama
1½ cups cooked wild rice
2 teaspoons grated orange rind
⅛ teaspoon salt
⅛ teaspoon pepper

Calories:	147	*Total fat:*	5.6 g
Protein:	2.4 g	*Saturated fat:*	0.8 g
Carbohydrates:	22.1 g	*Cholesterol:*	0
Fiber:	1.0 g	*Sodium:*	341 mg

✒ WILD RICE WITH SUGAR SNAPS (E)(V)

Sugar snaps, also called snap peas, are peas in a tender edible pod. They have tough strings on top and bottom that should be removed before cooking. If you can't find sugar snaps, use snow peas instead. For a heartier dish that could serve as an entrée, stir in cooked chickpeas at the same time you add the sugar snaps.

In a 1-quart saucepan, heat the oil over medium-high heat. Add the

2 tablespoons olive oil
2 tablespoons minced shallot
1 cup vegetable (or chicken) broth
½ cup wild rice
¼ teaspoon dried rosemary, crumbled
⅛ teaspoon dried thyme
⅛ teaspoon pepper
1 cup sugar snaps, tough strings removed

shallot and sauté until softened. Add the broth and bring to a boil. Stir in the rice, rosemary, thyme, and pepper. Return to a boil; reduce the heat and simmer, covered, 50 minutes. Stir in the sugar snaps and simmer, covered, 5 minutes longer.

SERVES 4

Calories:	160	*Total fat:*	8.0 g
Protein:	4.2 g	*Saturated fat:*	1.0 g
Carbohydrates:	18.2 g	*Cholesterol:*	0
Fiber:	2.9 g	*Sodium:*	238 mg

‿ PORCINI POLENTA (V)

POLENTA
2½ cups water
¾ cup yellow cornmeal
½ teaspoon salt

SAUCE
1½ tablespoons olive oil
¼ cup minced shallot
2 cups sliced mushrooms
1 cup sliced porcini or
 other wild mushroom
½ cup vegetable (or
 chicken) broth
2 tablespoons Madeira,
 Marsala, or sherry
⅛ teaspoon pepper
⅓ cup grated Parmesan
 (optional)

If porcini are not available, substitute any other flavorful wild mushroom, such as Portobello, oyster, or shiitake. You can use instant polenta, available in Italian specialty and gourmet stores, instead of making your own; the recipe will be just as good. Use enough to yield 2 cups.

For the polenta: In a 2-quart saucepan, stir together the water, cornmeal, and salt until there are no lumps. Bring to a boil over medium-high heat, stirring frequently. Reduce the heat to simmer and cook, stirring constantly, until the polenta pulls away from the side of the pan, about 30 minutes. Spoon the polenta into a serving bowl and make a well in the center. Cover with aluminum foil to keep warm.

For the sauce: Heat the oil in a

large skillet over medium heat. Add the shallot and cook, stirring, until softened. Stir in both types of mushroom and cook, stirring, until softened. Stir in the broth, Madeira, and pepper, and cook over high heat until the sauce is reduced by half, about 4 minutes. Stir in the Parmesan (if using).

Spoon the porcini sauce into the well in the polenta and serve immediately.

SERVES 4

Calories:	217	Total fat:	9.4 g
Protein:	6.3 g	Saturated fat:	2.2 g
Carbohydrates:	25.4 g	Cholesterol:	5.6 mg
Fiber:	2.1 g	Sodium:	495 mg

✿ BRUSSELS AND BARLEY (E)(V)

Both barley and brussels sprouts are wintery fare that make you feel full and happy (assuming you like them in the first place).

In a 2-quart saucepan, heat the oil over medium-high heat. Add the onion and cook, stirring, until softened. Add the brussels sprouts and water. Cover and cook, on high heat, 2 minutes, or until the brussels sprouts reach the desired doneness. Add the barley, salt, and pepper, and cook, over medium heat, until heated through.

SERVES 6

2 tablespoons vegetable oil
½ cup chopped onion
2 cups fresh or frozen brussels sprouts, quartered
3 tablespoons water
1½ cups cooked pearl barley
¼ teaspoon salt
¼ teaspoon pepper

Calories:	113	*Total fat:*	4.9 g
Protein:	3.3 g	*Saturated fat:*	0.7 g
Carbohydrates:	16.5 g	*Cholesterol:*	0
Fiber:	3.6 g	*Sodium:*	373 mg

❧ ONION BARLEY (E)(V)

---

2 tablespoons vegetable oil
1 cup chopped onion
1¼ cups water
½ cup hulled or hull-less
 barley, rinsed
1 cup frozen peas
½ teaspoon salt
¼ teaspoon pepper

You can substitute pearl barley for the hulled or hull-less; just reduce the cooking time to 40 minutes.

In a 1½-quart saucepan, heat the oil over medium-high heat. Add the onion and cook until golden-brown. Stir in the water and bring to a boil. Stir in the barley and return to a boil, reduce the heat, and simmer, covered, 1 hour. Stir in the peas, salt, and pepper. Simmer 10 minutes longer.

SERVES 4

Calories:	110	*Total fat:*	7.0 g
Protein:	2.7 g	*Saturated fat:*	1.0 g
Carbohydrates:	9.9 g	*Cholesterol:*	0
Fiber:	2.6 g	*Sodium:*	302 mg

❧ BARLEY PILAF (E)(V)

---

2 tablespoons vegetable oil
1 cup pearl barley, rinsed
½ cup finely chopped onion
2½ cups vegetable (or
 chicken) broth
2 tablespoons chopped
 parsley

The amount of salt you will need in this recipe will depend largely on the saltiness of the broth that you use. Season to taste before serving.

In a 2-quart saucepan, heat the oil over medium-high heat. Add the barley and cook, stirring, until it is browned and gives off a nutty aroma. Add the onion and sauté until softened.

Add the broth and bring to a boil over high heat. Reduce the heat to low and simmer, covered, 35 to 40 minutes, or until the liquid has been absorbed. Stir in the parsley, lemon juice, and pepper. Remove from the heat and let stand 5 minutes. Fluff with a fork.

SERVES 6 TO 8

1 tablespoon fresh lemon juice
¼ teaspoon pepper

Calories:	87	Total fat:	5.2 g
Protein:	1.6 g	Saturated fat:	0.7 g
Carbohydrates:	9.2 g	Cholesterol:	1.2 mg
Fiber:	1.0 g	Sodium:	360 mg

✎ QUINOA-SUNCHOKE PILAF (E)(Q)(V)

Sunchokes are also known as Jerusalem artichokes. Although they are not members of the artichoke family, they do taste remarkably like artichoke hearts when cooked. Raw sunchokes have a crispy texture and a slightly sweet flavor. Add them to salads for a delicious crunch.

Place the quinoa in a large bowl; fill with cold water. Pour into a strainer, then return the quinoa to the bowl and rinse 4 times more. Drain well.

Heat the oil in a 2-quart saucepan over medium-high heat. Add the rinsed quinoa and cook, stirring, until it crackles and pops, about 3

½ cup quinoa
2 tablespoons oil
½ cup chopped onion
1¼ cups vegetable (or chicken) broth
¾ cup cooked or canned (drained and rinsed) chickpeas
1 cup peeled, chopped sunchokes
½ cup fresh or frozen peas
¼ teaspoon pepper

to 5 minutes. Add the onion and cook, stirring, until the onion is soft.

Add the vegetable broth and bring to a boil over high heat. Add the chickpeas, sunchokes, peas, and pepper, and return to a boil. Reduce the heat and simmer, covered, 20 minutes. Fluff with a fork.

SERVES 6 TO 8

Calories:	172	*Total fat:*	7.7 g
Protein:	5.5 g	*Saturated fat:*	1.6 g
Carbohydrates:	22.8 g	*Cholesterol:*	5.5 mg
Fiber:	3.6 g	*Sodium:*	329 mg

‌ QUINOA-STUFFED TOMATOES (E)(V)

---

2 large tomatoes (or 4 small)
2 teaspoons olive oil
¼ cup chopped onion
1 clove garlic, minced
1 cup cooked quinoa
⅓ cup frozen peas
2 tablespoons chopped parsley
2 tablespoons grated Parmesan (optional)
⅛ teaspoon pepper

This is a perfect use for leftover quinoa. Add any chopped vegetables you like — zucchini, green or red bell pepper, mushrooms, celery, snow peas — and sauté with the onion.

Preheat the oven to 375° F.

Cut the tomatoes in half and scoop out the inner flesh and seeds; chop and set aside. If you're using small tomatoes, slice off the blossom end and scoop out the seeds and flesh; chop the flesh and set aside.

In a 1-quart saucepan, heat the oil over medium-high heat. Add the onion and garlic, and cook, stirring, until softened. Stir in the quinoa, peas, parsley, Parmesan (if using), pepper, and reserved chopped tomato. Spoon the quinoa mixture into the hollow tomato shells.

Place the filled tomatoes in an ungreased baking pan. Bake 15 minutes, or until heated through.

SERVES 4

Calories:	136	*Total fat:*	3.7 g
Protein:	4.6 g	*Saturated fat:*	0.9 g
Carbohydrates:	22.5 g	*Cholesterol:*	2.0 mg
Fiber:	4.1 g	*Sodium:*	223 mg

∾ CARROTY QUINOA (F)(Q)(V)

The cooked texture of quinoa is light and not at all what you expect from a grain. The cardamom serves as a perfect complement to the natural sweetness of the carrots.

Place the quinoa in a large bowl and fill with cold water. Pour into a strainer. Return the quinoa to the bowl and repeat 4 more times.

Heat the oil in a 2-quart saucepan over medium-high heat. Add the rinsed quinoa and carrot, and cook, stirring, until the quinoa smells nutty and starts to crackle and pop. Add the broth and cardamom. Bring to a boil. Reduce the heat and simmer, covered, 20 minutes. Fluff with a fork.

1 cup quinoa
1½ tablespoons vegetable oil
1 cup coarsely shredded carrot
2 cups vegetable (or chicken) broth
¼ teaspoon ground cardamom

SERVES 6 TO 8

Calories:	119	*Total fat:*	2.0 g
Protein:	3.8 g	*Saturated fat:*	0.3 g
Carbohydrates:	23.4 g	*Cholesterol:*	0
Fiber:	3.2 g	*Sodium:*	254 mg

∾ LEMON-HERB QUINOA   (E)(Q)(V)

1 cup quinoa
1½ tablespoons vegetable oil
2 cups water
¾ teaspoon dried marjoram
  or oregano
½ teaspoon dried thyme
¼ teaspoon dried rosemary,
  crumbled
3 tablespoons chopped
  parsley
2 tablespoons fresh lemon
  juice
¾ teaspoon salt
½ teaspoon grated lemon
  rind
¼ teaspoon pepper

You can substitute your favorite herbs for the ones called for here. And, for a fuller, richer flavor, use broth instead of water.

Place the quinoa in a large bowl; fill with cold water. Drain into a strainer and repeat the rinsing and draining 4 more times.

Over medium-high heat, heat the oil in a 2-quart saucepan. Add the rinsed quinoa and cook, stirring, until the quinoa makes cracking and popping noises, about 3 to 5 minutes. Stir in the water, marjoram, thyme, and rosemary. Bring to a boil, reduce the heat, and simmer, covered, 15 minutes.

Stir in the parsley, lemon juice, salt, lemon rind, and pepper. Simmer, covered, 5 minutes longer. Fluff with a fork.

SERVES 4 TO 6

Calories:	206	*Total fat:*	3.2 g
Protein:	6.0 g	*Saturated fat:*	1.2 g
Carbohydrates:	33.0 g	*Cholesterol:*	0
Fiber:	4.1 g	*Sodium:*	451 mg

❧ KASHA AND CAULIFLOWER (E)(V)

The interesting textures in this recipe — tender-crisp cauliflower, chewy mushrooms and raisins, and fluffy kasha — combine to make a delicious side dish. Cut the cauliflower into smallish pieces for the best result.

In a large skillet, heat the oil over medium-high heat. Add the cauliflower, mushrooms, and carrot. Cook, stirring, until the mushrooms are soft. Add the water and cook, stirring, until evaporated. Add the kasha and raisins, and cook, stirring, until heated through.

SERVES 4

2 tablespoons vegetable oil
1 cup cauliflower florets
1 cup sliced mushrooms
½ cup coarsely shredded carrot
3 tablespoons water
1 cup cooked kasha (roasted buckwheat)
⅓ cup golden raisins

Calories:	172	Total fat:	6.8 g
Protein:	4.4 g	Saturated fat:	1.0 g
Carbohydrates:	31.0 g	Cholesterol:	0
Fiber:	4.2 g	Sodium:	188 mg

❧ KASHA AND BOW TIES (E)(Q)(V)

For an even more nutritious version of this recipe, omit the chopped cashews. You will have a lower total fat content, as well as less sodium and calories, and the dish will be just as delicious.

In a large skillet, heat the oil over medium-high heat. Add the cabbage and onion, and cook, stirring, until softened. Add the kasha, bow ties,

1 tablespoon vegetable oil
1 cup chopped cabbage
½ cup chopped onion
1 cup cooked kasha (roasted buckwheat)
¾ cup cooked small bow ties or other small pasta
¼ teaspoon salt
¼ teaspoon pepper
⅓ cup chopped cashews

¼ teaspoon salt
¼ teaspoon pepper
⅓ cup chopped cashews

salt, and pepper. Cook, stirring, until heated through. Stir in the cashews.

SERVES 4

Calories:	193	*Total fat:*	9.0 g
Protein:	5.5 g	*Saturated fat:*	1.6 g
Carbohydrates:	22.6 g	*Cholesterol:*	0
Fiber:	3.3 g	*Sodium:*	205 mg

❧ STEWED CHICKPEAS AND OKRA (E)(V)

2 tablespoons vegetable oil
1 cup sliced fresh okra (½-inch pieces)
½ cup chopped onion
½ cup chopped green bell pepper
1 can (14½ ounces) whole peeled tomatoes
¼ teaspoon dried oregano
¼ teaspoon ground cumin
⅛ teaspoon salt
⅛ teaspoon ground red pepper
1 cup cooked or canned (drained and rinsed) chickpeas

I'm not usually a great fan of okra, because of its sliminess. This delicious dish, however, has a perfectly pleasing consistency.

In a 2-quart saucepan, heat the oil over medium-high heat. Add the okra, onion, and bell pepper, and cook, stirring, until softened. Stir in the tomatoes, breaking them up with the back of a spoon. Add the oregano, cumin, salt, and red pepper. Stir in the chickpeas and bring to a boil; lower the heat and simmer, uncovered, 10 minutes.

SERVES 6 TO 8

Calories:	117	*Total fat:*	5.6 g
Protein:	3.9 g	*Saturated fat:*	0.4 g
Carbohydrates:	14.5 g	*Cholesterol:*	0
Fiber:	3.7 g	*Sodium:*	233 mg

~ DILLED LIMA BEANS (E)(Q)(V)

You can use baby or Fordhook lima beans, according to your preference. Stirring in 1 cup of corn kernels when you add the beans will transform this dish into succotash.

In a medium skillet, over medium heat, melt the margarine. Add the celery and garlic, and cook, stirring, until tender-crisp. Add the lima beans, dill, salt, and pepper. Cook, stirring occasionally, 5 minutes, or until heated through.

SERVES 4

1 tablespoon margarine or vegetable oil
1 cup sliced celery
1 clove garlic, minced
1 package (10 ounces) frozen lima beans
2 tablespoons fresh snipped dill
½ teaspoon salt
⅛ teaspoon pepper

Calories:	134	*Total fat:*	2.8 g
Protein:	5.0 g	*Saturated fat:*	0.6 g
Carbohydrates:	15.2 g	*Cholesterol:*	0
Fiber:	6.5 g	*Sodium:*	316 mg

~ BRAISED FRENCH LENTILS (E)(V)

Le Puy, or French, lentils are smaller and darker brown than the lentils commonly sold in supermarkets. They are prized as the most flavorful of all lentils. They cook a little more quickly than the green or regular brown variety. If French lentils are not available, substitute the supermarket variety, but be aware that they may need extra cooking time.

Bring the water to a boil in a 1½ quart saucepan over medium-high

3 cups water
1 cup French lentils, rinsed
1½ tablespoons olive oil
1 cup chopped onion
½ cup chopped celery
½ cup chopped carrot
1 clove garlic, minced
½ teaspoon dried rosemary, crumbled
¼ teaspoon dried thyme
⅓ cup vegetable (or chicken) broth
3 tablespoons chopped

parsley
1 cup cauliflower florets
¾ teaspoon salt
⅛ teaspoon pepper

heat. Add the lentils; reduce the heat and simmer 30 minutes, or until just tender; drain.

In a 2-quart saucepan, heat the oil over medium-high heat. Add the onion, celery, carrot, garlic, rosemary, and thyme; sauté until the onion is soft. Add the drained lentils, broth, and parsley, and return to a boil. Simmer 5 minutes. Add the cauliflower, salt, and pepper, and simmer 5 minutes longer, or until the cauliflower is tender.

SERVES 6 TO 8

Calories:	158	*Total fat:*	3.9 g
Protein:	9.9 g	*Saturated fat:*	0.5 g
Carbohydrates:	22.5 g	*Cholesterol:*	0
Fiber:	5.2 g	*Sodium:*	386 mg

✍ BOSTON BAKED BEANS (V)

---

3 cups cooked small white
   beans
½ cup water
¼ cup finely chopped onion
3 tablespoons firmly
   packed light or dark
   brown sugar
3 tablespoons molasses
2 teaspoons dry mustard
1 teaspoon salt
¼ teaspoon ground cloves
¼ teaspoon pepper

These beans are sweet, but not as gooey as the ones that come in a can.

Preheat the oven to 275° F.

Place the beans in a 1½-quart ovenproof casserole. Add the water, onion, brown sugar, molasses, mustard, salt, cloves, and pepper. Stir until combined. Cover tightly and bake 6 hours, or until the beans are tender and coated with sauce. Check the beans during cooking time and add water as necessary to keep from sticking.

SERVES 6

Calories:	167	Total fat:	0.6 g
Protein:	8.1 g	Saturated fat:	0.2 g
Carbohydrates:	33.5 g	Cholesterol:	0
Fiber:	7.2 g	Sodium:	361 mg

✁ BARBECUED BEANS (E)(V)

Pickapeppa Sauce, a Jamaican condiment, is a combination of tomato, onion, mango, raisin, tamarind, peppers, and spices. If you can't find it in your supermarket, steak sauce will be a fine substitute.

Preheat the oven to 350° F.

Heat the oil in a medium skillet over medium-high heat. Add the onion, bell pepper, and garlic; cook, stirring, until softened. Remove from the heat and stir in the ketchup, orange juice, brown sugar, molasses, Pickapeppa Sauce, mustard, Worcestershire sauce, salt, and red pepper. Stir in both types of bean. Spoon into a 1½-quart casserole. Bake, covered, 1 hour.

SERVES 6 TO 8

2 tablespoons vegetable oil
1 cup chopped onion
1 cup chopped green bell pepper
1 clove garlic, minced
½ cup ketchup
¼ cup orange juice
3 tablespoons firmly packed light or dark brown sugar
1 tablespoon dark molasses
1 tablespoon Pickapeppa Sauce
1 tablespoon prepared brown mustard
2 teaspoons Worcestershire sauce
¼ teaspoon salt
⅛ teaspoon ground red pepper (optional)
1¾ cups cooked or canned (drained and rinsed) small white beans
1¼ cups cooked or canned (drained and rinsed) red kidney beans

Calories:	232	Total fat:	6.8 g
Carbohydrates:	8.9 g	Saturated fat:	0.6 g
Protein:	39.3 g	Cholesterol:	0
Fiber:	8.2 g	Sodium:	457 mg

᷒ REFRIED BEANS (E)(V)

1 tablespoon vegetable oil
⅓ cup chopped onion
1 clove garlic, minced
1¾ cups cooked or canned
  (drained and rinsed)
  pinto beans
¼ cup water
¼ teaspoon ground cumin
¼ teaspoon ground red
  pepper (or to taste)
¼ teaspoon salt

I find cooked rice and refried beans, plus salsa and a simple lettuce-and-tomato salad, to be an extremely satisfying meal. This is a very quick dish if you are starting out with canned beans.

In a 1½-quart saucepan, heat the oil over medium-high heat. Add the onion and garlic, and cook, stirring, until softened. Add the beans, water, cumin, red pepper, and salt. Stir, mashing about half the beans with the back of a spoon. Cook, stirring, until heated through.

SERVES 4

Calories:	135	*Total fat:*	5.2 g
Protein:	4.7 g	*Saturated fat:*	0.8 g
Carbohydrates:	18.9 g	*Cholesterol:*	0
Fiber:	4.7 g	*Sodium:*	253 mg

᷒ STUFFED ACORN SQUASH WITH BUTTER BEANS AND CHESTNUTS

2 small acorn squash,
  halved and seeded
1½ tablespoons oil
1 tablespoon chopped shallot
1½ cups cooked or canned
  (rinsed and drained) butter beans
½ cup chopped cooked
  chestnuts

The rich flavor of the butter beans combines perfectly with the texture and taste of the squash.

Preheat the oven to 375° F. Place the squash, cut side up, in a baking pan filled with 1 inch of water. Bake 45 minutes, or until tender. Or pierce the skin with a fork and microwave on high (100 percent) power 8 to 10 minutes, or until tender, turning over once. Let stand 4 minutes. Cut in half and discard the strings and seeds.

In a large skillet, heat the oil over medium-high heat. Add the shallot and cook, stirring, until softened. Add the beans, chestnuts, parsley, milk, and sherry. Cook, stirring, until the beans are glazed and heated through. Place a quarter of the bean mixture into each baked squash half.

SERVES 4

2 tablespoons chopped parsley
2 tablespoons milk
1 tablespoon cream sherry

Calories:	239	Total fat:	6.0 g
Protein:	6.5 g	Saturated fat:	0.9 g
Carbohydrates:	42.1 g	Cholesterol:	1.0 mg
Fiber:	11.6 g	Sodium:	221 mg

❧ HONEY-GLAZED SQUASH, GREENS, AND BEANS (E)(V)

I use spinach as my greens in this dish; you can use any green that you like, such as kale or Swiss chard. I prefer butternut squash for this recipe, because it's easy to peel, but you can substitute any other winter squash. You can substitute cooked Fordhook lima beans or cannellini for the butter beans. If you're not fond of anise flavoring, just omit it. Serve this dish with brown rice on the side.

To cook the squash in the microwave, place it in a microwave-safe bowl. Cover with waxed paper and cook on high (100 percent) power

2 cups butternut squash, cut into 1-inch cubes
⅓ cup sliced almonds
2 tablespoons vegetable oil
¼ teaspoon anise seeds
3 cups firmly packed, thoroughly rinsed, coarsely chopped greens
1½ cups cooked or canned (drained and rinsed) butter beans
2 tablespoons honey
¼ teaspoon grated orange rind
¼ teaspoon salt

3 minutes, or until softened; set aside. To prepare in a conventional oven, halve and seed the squash. Bake at 350° F until tender, about 40 minutes, and then cut into 1-inch cubes.

Place the almonds in a large skillet. Cook over medium heat until browned. Remove from the skillet and set aside. Add the oil and anise seeds to the skillet and cook 30 seconds, to bring out the flavor. Add the reserved squash and the greens, beans, honey, orange rind, and salt. Cook, stirring, over high heat, until any liquid in the bottom of the skillet evaporates.

Stir in the browned almonds and serve.

SERVES 6

Calories:	181	*Total fat:*	6.9 g
Protein:	7.4 g	*Saturated fat:*	1.6 g
Carbohydrates:	26.3 g	*Cholesterol:*	0
Fiber:	8.4 g	*Sodium:*	373 mg

✺ CHILI POTATO WEDGES   (E)(V)

---

2 unpeeled baking potatoes, well rinsed
1 tablespoon vegetable oil
1 tablespoon chili powder
2 cloves garlic, minced
½ teaspoon dried oregano
½ teaspoon salt
¼ teaspoon ground cumin
¼ teaspoon ground red pepper (optional)

In this recipe I don't peel the potatoes. This makes them more attractive and allows them to retain all the nutrients found in the skin. The red pepper adds a definite tingle, but it's optional.

Preheat the oven to 400° F.

Cut each potato in half lengthwise. Cut each half into wedges that measure about ½-inch thick on the skin side; you'll get 8 to 12 wedges per potato. Set aside.

In a small bowl, stir together the oil, chili powder, garlic, oregano, salt, cumin, and red pepper (if using). Place the potato wedges in an ungreased 9 × 13 × 2-inch baking pan and drizzle the chili mixture over them. Toss to completely coat each wedge. Bake 45 minutes, or until the wedges are crispy on the outside and tender on the inside.

SERVES 4

Calories:	148	Total fat:	3.8 g
Protein:	2.7 g	Saturated fat:	0.6 g
Carbohydrates:	27.0 g	Cholesterol:	0
Fiber:	2.8 g	Sodium:	294 mg

HERBED SCALLOPED POTATOES (V)

The herbs dress up an old family favorite. I make this version with vegetable broth, so that it's totally vegetarian. The dish will be quite soupy when it first comes out of the oven, but the sauce thickens on standing.

Preheat the oven to 375° F. Grease an 8 × 8 × 2-inch baking pan.

Cook the potatoes in boiling water 5 minutes, or until just tender; drain.

Toss together the onion, parsley, flour, salt, thyme, and pepper; set aside.

Layer a third of the potatoes in the prepared baking dish. Sprinkle

4 cups peeled and thinly sliced boiling potatoes
1 cup thinly sliced onion
¼ cup chopped parsley
2 tablespoons all-purpose flour
½ teaspoon salt
¼ teaspoon dried thyme
¼ teaspoon pepper
1 cup hot vegetable (or chicken) broth
1½ teaspoons olive oil

with half of the herb mixture. Repeat layering, with the next third of the potatoes and the remaining herb mixture. Top with the remaining potatoes. Pour the broth over the potatoes and drizzle with the oil.

Cover and bake 30 minutes; uncover and bake 30 minutes longer, or until the potatoes are tender. Let stand 10 to 15 minutes before serving.

SERVES 4 TO 6

Calories:	184	*Total fat:*	2.4 g
Protein:	4.0 g	*Saturated fat:*	0.4 g
Carbohydrates:	37.5 g	*Cholesterol:*	0
Fiber:	3.0 g	*Sodium:*	322 mg

✎ STUFFED BAKED POTATOES

2 large baking potatoes, well rinsed and pierced with a fork
¼ cup cottage cheese
2 tablespoons buttermilk or unflavored yogurt
3 tablespoons chopped scallion
2 tablespoons chopped parsley
¾ teaspoon salt
⅛ teaspoon pepper

It's easy to make delicious stuffed potatoes if you use oodles of butter or cream cheese. The challenge is to make yummy low-fat stuffed potatoes — and these are yummy indeed. You can transform these into twice-baked potatoes by baking them for 15 to 20 minutes, or until browned on top, after you've stuffed them.

Preheat the oven to 350° F.

Bake the potatoes 1 hour, or until soft.

Place the cottage cheese, buttermilk, scallion, parsley, salt, and pepper in the workbowl of a food processor fitted with a steel blade. Cover and process until smooth.

Slice the baked potatoes in half lengthwise.

Scoop out most of the potato, leaving the shell ⅛ inch thick; set aside the shells. Place the potato in the food processor with the cottage cheese mixture. Cover and process, pulsing, until just blended (if you process until smooth, the potatoes become overworked and sticky). Spoon into the potato shells.

SERVES 4

Calories:	130	*Total fat:*	0.8 g
Protein:	4.5 g	*Saturated fat:*	0.5 g
Carbohydrates:	26.6 g	*Cholesterol:*	2.4 mg
Fiber:	2.6 g	*Sodium:*	474 mg

~ ORIENTAL GRILLED VEGETABLES (E)(Q)(V)

Although yellow squash is not particularly Oriental, it goes well with the other vegetables in this recipe. If you can't find yellow squash, substitute zucchini.

Put the coriander seeds into a medium bowl and crush, using the back of a spoon. Add the soy sauce, sherry, honey, ginger, oil, and garlic. Add the vegetables and toss. Let stand 15 minutes.

Preheat the grill or broiler. String the vegetables onto 4 six-inch skewers. Cook 4 inches from the heat source, 3 minutes per side, brushing once with the marinade.

SERVES 4

1 tablespoon coriander seeds
2 tablespoons soy sauce
2 tablespoons dry sherry
1 tablespoon honey
2 teaspoons minced fresh ginger
1 teaspoon chili or sesame oil
2 cloves garlic
1 large yellow squash, cut into ½-inch slices
1 large onion, cut into 8 wedges
1 red bell pepper, cut into 8 pieces
1 green bell pepper, cut into 8 pieces

Calories:	77	*Total fat:*	2.0 g
Protein:	2.4 g	*Saturated fat:*	0.3 g
Carbohydrates:	13.5 g	*Cholesterol:*	0
Fiber:	2.4 g	*Sodium:*	519 mg

❧ SAUTÉED BABY SQUASH AND FENNEL (E)(Q)(V)

1 tablespoon olive oil
2 cups baby squash (or any combination of miniature zucchini, yellow squash, or pattypan squash)
1 cup sliced fennel
¾ cup sliced red onion
1 tablespoon fresh lemon juice
½ teaspoon fennel seeds (optional)
½ teaspoon salt
⅛ teaspoon pepper

You can use any combination of baby squash that you find in your market for this recipe. If the miniature squash are not available, use sliced zucchini and/or yellow squash.

In a medium or large skillet, heat the oil over medium-high heat. Add the squash, fennel, and onion. Cook, stirring, until the onion is softened. Add the lemon juice, fennel seeds (if using), salt, and pepper, and cook, covered, about 3 to 5 minutes, or until the vegetables are tender.

SERVES 4

Calories:	80	*Total fat:*	4.1 g
Protein:	1.1 g	*Saturated fat:*	0.8 g
Carbohydrates:	12.8 g	*Cholesterol:*	0
Fiber:	3.2 g	*Sodium:*	270 mg

✺ ORANGE PUREE OF SQUASH AND CARROTS (E)(Q)(V)

This puree is orange in both color and flavor. The almonds bring a welcome texture, not to mention great flavor, to the vegetable combination. You can substitute other sweet winter squash, such as buttercup, for the butternut.

Preheat the oven to 350° F. Place the almonds in a baking pan and bake 10 minutes, or until lightly browned; set aside.

Cook the carrots over high heat, in boiling water, 10 minutes. Add the squash and cook 10 minutes longer, or until the vegetables are tender; drain.

Place the vegetables, orange juice, brown sugar, orange rind, and salt into the workbowl of a food processor fitted with a steel blade. Cover and process until smooth. Spoon the puree into a bowl. Stir in all but 1 tablespoon of the toasted almonds. Garnish with the remaining almonds.

SERVES 4

⅓ cup sliced almonds
2 cups carrots, cut into 2-inch pieces
2 cups cubed butternut squash
½ cup orange juice
¼ cup firmly packed light or dark brown sugar
1 tablespoon grated orange rind
¼ teaspoon salt

Calories:	212	Total fat:	6.3 g
Protein:	3.9 g	Saturated fat:	0.6 g
Carbohydrates:	39.0 g	Cholesterol:	0
Fiber:	6.7 g	Sodium:	141 mg

❧ SHERRIED SALSIFY AND CARROTS (E)(Q)(V)

2 cups julienned salsify
2 cups julienned carrot
1 tablespoon margarine
3 tablespoons firmly
  packed light or dark
  brown sugar
2 tablespoons dry sherry
1 teaspoon fresh lemon
  juice
¼ teaspoon ground nutmeg
⅛ teaspoon salt

Salsify is a long, thin root plant, black or white in color. It's also known as the oyster plant, although it neither looks like nor tastes like an oyster. Its flavor is extremely delicate, and its texture is somewhat like that of parsnip (which you can substitute for salsify, if it is not available). Salsify must be peeled before cooking.

Cook the salsify and carrot in boiling water, over high heat, 5 minutes. Drain.

In a medium or large skillet, melt the margarine over medium-high heat. Stir in the sugar, sherry, lemon juice, nutmeg, and salt. Add the drained salsify and carrot. Cook, stirring, until the vegetables absorb the sherry mixture.

SERVES 4 TO 6

Calories:	143	*Total fat:*	3.1 g
Protein:	2.4 g	*Saturated fat:*	0.6 g
Carbohydrates:	26.1 g	*Cholesterol:*	0
Fiber:	3.3 g	*Sodium:*	135 mg

❧ FRUITY NOODLE PUDDING (E)

In addition to being a lovely side dish, this pudding makes an excellent brunch or lunch entrée, or even a dessert. You can use pineapple cottage cheese instead of the plain for an even fruitier result.

Preheat the oven to 350° F. Lightly grease a 9-inch-square baking pan.

Cook the noodles in a large pot of salted boiling water, according to package directions; drain and set aside.

In a large bowl, stir together the cottage cheese, applesauce, yogurt, sugar, cinnamon, and salt. Stir in the egg whites. Stir in the drained noodles and the raisins. Spoon into the prepared pan.

Bake 50 minutes, or until the top is browned. Let stand 10 minutes before serving. Cut into 12 portions.

SERVES 12

1 package (12 ounces) no-yolk egg noodles
1 container (16 ounces) cottage cheese
1 cup chunky applesauce
1 container (8 ounces) vanilla yogurt
⅓ cup sugar
1½ teaspoons ground cinnamon
1 teaspoon salt
2 egg whites
1 cup golden raisins

Calories:	236	*Total fat:*	2.3 g
Protein:	10.2 g	*Saturated fat:*	1.3 g
Carbohydrates:	44.4 g	*Cholesterol:*	6.5 mg
Fiber:	2.2 g	*Sodium:*	354 mg

Salads

WILD RICE, YELLOW PEPPER, AND BLACK-EYED PEA SALAD (E)(V)

2 cups cooked wild rice
1¾ cups cooked or canned (drained and rinsed) black-eyed peas
1 cup chopped yellow (or green) bell pepper
1 cup chopped jicama
1 cup chopped zucchini
12 cherry tomatoes, halved
¼ cup chopped parsley
3 tablespoons olive oil
3 tablespoons fresh lemon juice
2 tablespoons vegetable oil
1 tablespoon red wine vinegar
1 tablespoon Dijon mustard
1 small clove garlic, minced
1 teaspoon salt
½ teaspoon dried rosemary, crumbled
½ teaspoon pepper

This is an interesting combination of wild rice, usually considered very "upper crusty," and black-eyed peas, supposedly "poor man's" food. If the wild rice proves too costly for your budget, substitute white or brown rice.

In a large bowl, combine the wild rice, black-eyed peas, bell pepper, jicama, zucchini, tomatoes, and parsley.

In a small bowl, stir together the olive oil, lemon juice, vegetable oil, vinegar, mustard, garlic, salt, rosemary, and pepper. Pour over the salad and toss.

SERVES 12

Calories:	117	Total fat:	6.0 g
Protein:	3.6 g	Saturated fat:	0.7 g
Carbohydrates:	13.2 g	Cholesterol:	0
Fiber:	3.3 g	Sodium:	265 mg

❧ HEARTS OF PALM AND CHICKPEA SALAD (E)(Q)(V)

Hearts of palm are the interior of small cabbage palm trees. Grown in Florida and other tropical regions, they are usually available canned and packed in water, or marinated. If you have trouble finding the marinated hearts, use the water-packed ones and add vinegar to taste. For a heartier meal, stir in drained tuna.

In a large bowl, toss together the chickpeas, tomato, hearts of palm, olives, scallion, and parsley. Add the oil, lemon juice, garlic, oregano, and pepper, and toss until thoroughly combined.

SERVES 6

1¼ cups cooked or canned (drained and rinsed) chickpeas
1 cup coarsely chopped tomato
1 jar (6 ounces) marinated hearts of palm, drained, or ¾ cup hearts of palm, cut into 1-inch pieces
20 pitted black olives, halved
¼ cup chopped scallion
¼ cup chopped parsley
2 tablespoons olive oil
2 tablespoons fresh lemon juice
½ clove garlic, minced
¼ teaspoon dried oregano
¼ teaspoon pepper

Calories:	137	*Total fat:*	9.0 g
Protein:	3.9 g	*Saturated fat:*	1.3 g
Carbohydrates:	13.1 g	*Cholesterol:*	0
Fiber:	3.9 g	*Sodium:*	175 mg

❧ PINTO SALAD WITH HONEY-GINGER DRESSING (E)(V)

The raisins give this salad a surprising taste and texture. Toss it at the last minute to prevent all the vegetables from becoming shocking pink from the beets.

2 tablespoons cider vinegar
1 tablespoon vegetable oil
2 teaspoons honey
½ teaspoon dry mustard
½ teaspoon salt

¼ teaspoon ground ginger

⅛ teaspoon pepper

1¾ cups cooked or canned (drained and rinsed) pinto beans

½ cup cooked, julienned beets

½ cup chopped zucchini

½ cup chopped red bell pepper

¼ cup raisins

3 tablespoons finely chopped red onion

In a small bowl, stir together the vinegar, oil, honey, mustard, salt, ginger, and pepper.

In a large bowl, toss together the beans, beets, zucchini, bell pepper, raisins, and onion.

Pour the dressing over the salad and toss to combine.

SERVES 4 TO 6

Calories:	190	*Total fat:*	2.7 g
Protein:	7.0 g	*Saturated fat:*	0.6 g
Carbohydrates:	33.0 g	*Cholesterol:*	0
Fiber:	10.0 g	*Sodium:*	336 mg

❧ RED AND WHITE KIDNEY BEAN SALAD WITH ARTICHOKES AND OLIVES (E)(V)

¼ cup pignoli (pine nuts)

1¼ cups cooked or canned (drained and rinsed) red kidney beans

1¼ cups cooked or canned (drained and rinsed) white kidney beans

1 cup chopped tomato

1 jar (6 ounces) marinated artichokes, drained and coarsely chopped

½ cup chopped green bell pepper

½ cup sliced black olives

⅓ cup chopped red onion

Using marinated artichokes and olives adds lots of salt to this salad; sodium watchers may want to omit one or both of these items.

Place the pignoli in a skillet and cook over medium-low heat, shaking the pan, until the nuts are browned.

In a large bowl, combine both types of kidney bean and the tomato, artichokes, bell pepper, olives, onion, pignoli, and parsley. Add the vinegar, oil, and pepper. Toss to combine.

SERVES 6 TO 8

¼ cup chopped parsley
1 tablespoon + 1 tea-
  spoon red wine vinegar
1 tablespoon olive oil
¼ teaspoon pepper

Calories:	202	Total fat:	17.2 g
Protein:	8.6 g	Saturated fat:	2.6 g
Carbohydrates:	21.1 g	Cholesterol:	0
Fiber:	10.5 g	Sodium:	558 mg

～ MARINATED CHICKPEA SALAD (E)(V)

I tasted a salad like this in a diner close to my parents' home. I loved it so much that I attempted to duplicate it right away. This is the result, and I think it's just about as delicious as the original.

In a large bowl, combine the chickpeas, corn, onion, carrot, and celery.

In a small bowl, stir together the oil, both vinegars, garlic, mustard, salt, sugar, paprika, and pepper. Pour the dressing over the salad and toss until combined. Let stand at least 1 hour before serving.

SERVES 8

1½ cups cooked or canned
  (drained and rinsed)
  chickpeas
1 can (8¾ ounces) corn
  kernels (1 cup), drained
⅓ cup chopped red onion
¼ cup shredded carrot
¼ cup finely chopped celery
3 tablespoons vegetable oil
1½ tablespoons cider vinegar
1 tablespoon distilled
  white vinegar
1 clove garlic, minced
½ teaspoon dry mustard
½ teaspoon salt
¼ teaspoon sugar
¼ teaspoon paprika
¼ teaspoon pepper

Calories:	117	Total fat:	3.9 g
Protein:	5.5 g	Saturated fat:	0.5 g
Carbohydrates:	16.5 g	Cholesterol:	0
Fiber:	6.1 g	Sodium:	293 mg

❧ BROCCOLI-BEAN SALAD WITH BALSAMIC DRESSING (E)(V)

2 cups blanched broccoli florets
1½ cups cooked or canned (drained and rinsed) cannellini
¾ cup sliced zucchini
½ cup chopped red bell pepper
3 tablespoons chopped red onion
1½ tablespoons olive oil
1 tablespoon balsamic vinegar
1 teaspoon honey mustard
½ teaspoon salt
⅛ teaspoon pepper

The honey mustard in this salad dressing emphasizes the naturally sweet flavor of the balsamic vinegar. Balsamic vinegar is very much like wine — cheap ones have a coarse, sour flavor, better ones a well-rounded bouquet.

In a large bowl, toss together the broccoli, cannellini, zucchini, bell pepper, and onion.

In a small bowl, mix the oil, vinegar, mustard, salt, and pepper. Pour the dressing over the salad and toss to combine. Let stand at least 10 minutes to let the flavors meld.

SERVES 4 TO 6

Calories:	163	*Total fat:*	5.8 g
Protein:	8.7 g	*Saturated fat:*	0.9 g
Carbohydrates:	22.4 g	*Cholesterol:*	0
Fiber:	7.4 g	*Sodium:*	314 mg

❧ ASPARAGUS AND CHICKPEA SALAD WITH GREEN PEPPERCORN DRESSING (E)(Q)(V)

2 cups blanched asparagus tips or whole asparagus, cut into 1½-inch pieces
1¼ cups cooked or canned (drained and rinsed) chickpeas

Green peppercorns are fresh peppercorns preserved in brine. You can buy them in most gourmet stores. I sometimes serve this salad on a bed of red and green lettuce leaves for an attractive appetizer.

In a large bowl, toss together the asparagus tips, chickpeas, fennel, olives, and scallion.

Place the oil, mayonnaise, mustard, peppercorns, vinegar, Worcestershire sauce, salt, and pepper in a blender container. Cover and blend until the peppercorns are finely chopped. Pour over the salad and toss.

SERVES 6

½ cup finely chopped fennel or celery

⅓ cup sliced black olives

¼ cup chopped scallion

1½ tablespoons olive oil

1 tablespoon mayonnaise

2 teaspoons Dijon mustard

1 teaspoon green peppercorns

1 teaspoon red wine vinegar

¼ teaspoon white wine Worcestershire sauce

¼ teaspoon salt

⅛ teaspoon pepper

Calories:	115	Total fat:	6.1 g
Protein:	4.7 g	Saturated fat:	0.9 g
Carbohydrates:	11.7 g	Cholesterol:	1.4 mg
Fiber:	2.8 g	Sodium:	345 mg

ORIENTAL ADZUKI BEAN SALAD (E)(V)

I usually find the flavor of sesame oil too strong for salad dressing, but the little bit called for here simply rounds out the flavor. If you like, you can substitute hot sesame oil for a spicy salad.

In a large bowl, stir together the beans, sprouts, radish, snow peas, and scallion.

In a small bowl, mix the oil, vinegar, soy sauce, *mirin*, sesame oil, garlic, and ginger. Pour over the salad and toss.

SERVES 4

1 cup cooked or canned (drained and rinsed) adzuki beans

1 cup bean sprouts

½ cup sliced radish

½ cup chopped snow peas

¼ cup sliced scallion

1 tablespoon vegetable oil

2 teaspoons cider vinegar

2 teaspoons soy sauce

2 teaspoons *mirin* or dry sherry

1 teaspoon sesame oil

1 clove garlic, minced

¼ teaspoon minced fresh ginger

Calories:	130	Total fat:	5.0 g
Protein:	6.0 g	Saturated fat:	0.5 g
Carbohydrates:	16.1 g	Cholesterol:	0
Fiber:	5.3 g	Sodium:	279 mg

∾ SPANISH WHITE BEAN–CAULIFLOWER SALAD (E)(Q)(V)

1 medium head cauliflower
1 cup cooked or canned (drained and rinsed) cannellini
12 pitted or stuffed green olives, sliced
1 jar (4 ounces) pimentos, drained and chopped
2 tablespoons olive oil
2 tablespoons red wine vinegar
2 tablespoons minced shallot
1 tablespoon chopped sweet pickle
¾ teaspoon salt
¼ teaspoon pepper

Since Spain is one of the world's leading producers of olives and olive oil, any Spanish salad would naturally include them. This one is no exception.

Cut the cauliflower into bite-size pieces. Cook in boiling water until tender; rinse under cold water and drain. Place in a large bowl; add the cannellini, olives, and pimentos.

In a small bowl, stir together the oil, vinegar, shallot, pickle, salt, and pepper. Pour over the salad and toss until combined.

SERVES 6 TO 8

Calories:	120	Total fat:	6.2 g
Protein:	4.9 g	Saturated fat:	0.8 g
Carbohydrates:	14.0 g	Cholesterol:	0
Fiber:	6.3 g	Sodium:	269 mg

❧ TOMATO, VIDALIA ONION, AND CHICKPEA SALAD (E)(Q)(V)

Vidalia onions are very mild in flavor. If you can't find them, use Spanish or Bermuda onions instead. Fresh basil, which makes this salad special, is sometimes hard to find; 1 teaspoon dried basil is an acceptable substitute. Don't bother making the recipe with anything but great tomatoes.

In a large bowl, toss the tomato, chickpeas, onion, and basil.

In a small bowl, stir together the vinegar, oil, mustard, salt, and pepper. Pour over the salad and toss. Add the Parmesan (if using) and toss again.

SERVES 6 TO 8

3 cups coarsely chopped tomato
1 cup cooked or canned (drained and rinsed) chickpeas
¾ cup coarsely chopped Vidalia onion
¼ cup chopped fresh basil
1½ tablespoons red wine vinegar
1 tablespoon olive oil
1 teaspoon Dijon mustard
¼ teaspoon salt
¼ teaspoon pepper
2 tablespoons grated Parmesan (optional)

Calories:	100	*Total fat:*	3.2 g
Protein:	3.5 g	*Saturated fat:*	0.4 g
Carbohydrates:	15.5 g	*Cholesterol:*	0
Fiber:	3.6 g	*Sodium:*	273 mg

❧ TRADITIONAL THREE-BEAN SALAD (E)(Q)(V)

Which three kinds of beans you use in this recipe is up to you. You can try wax beans or even a third type of dried bean instead of the fresh green beans. The important point is that this is a marinated salad, so be sure to let it stand at least the suggested 30 minutes.

1 cup cooked or canned (drained and rinsed) red kidney beans
1 cup cooked or canned (drained and rinsed) chickpeas
1 cup fresh green beans, cut into 1½-inch pieces

and blanched, *or* 1 cup
frozen cut green beans,
thawed
½ cup sliced celery
⅓ cup chopped onion
3 tablespoons vegetable oil
1½ tablespoons cider vinegar
½ teaspoon dry mustard
1 clove garlic, minced
¼ teaspoon Worcestershire
sauce
¼ teaspoon celery seed
¼ teaspoon sugar
⅛ teaspoon pepper

In a large bowl, toss together all
the beans and the celery and onion.

In a small bowl, combine the oil,
vinegar, mustard, garlic, Worcestershire sauce, celery seed, sugar, and
pepper. Pour the dressing over the
salad and toss until combined. Let
marinate at least 30 minutes before
serving.

SERVES 4 TO 6

Calories:	232	*Total fat:*	11.5 g
Protein:	8.5 g	*Saturated fat:*	1.5 g
Carbohydrates:	25.6 g	*Cholesterol:*	0
Fiber:	7.8 g	*Sodium:*	278 mg

❧ KIDNEY BEAN SALAD WITH AVOCADO DRESSING (E)(Q)

1½ cups cooked or canned
(drained and rinsed) kid-
ney beans
¾ cup sliced celery
½ cup chopped onion
½ cup chopped cucumber
⅓ cup chopped green bell
pepper
2 tablespoons chopped
fresh coriander (cilantro)
(optional)
½ small ripe avocado
(about ½ cup diced)
⅓ cup buttermilk
2 tablespoons fresh lime
juice

*Because avocado turns brown
when exposed to air, I recommend
that you prepare the dressing for
the salad just before serving time.*

In a large bowl, toss together the
beans, celery, onion, cucumber, bell
pepper, and coriander.

Put the avocado, buttermilk, lime
juice, garlic, salt, and pepper into a
blender container. Cover and blend
until smooth. Pour over the salad
and toss to combine.

SERVES 4 TO 6

1 clove garlic
¼ teaspoon salt
¼ teaspoon pepper

Calories:	152	*Total fat:*	4.6 g
Protein:	7.6 g	*Saturated fat:*	0.8 g
Carbohydrates:	22.5 g	*Cholesterol:*	0.7 mg
Fiber:	9.3 g	*Sodium:*	313 mg

✍ LENTIL-POTATO SALAD (E)

I'm not usually fond of potato salad, but this one, which can be served warm or cold, is fantastic. The recipe serves four, but not too generously. If you love potato salad, make a double batch.

Cook the potatoes in boiling water over medium heat, uncovered, 20 minutes, or until tender when pierced with a fork. Drain and, if desired, peel. Slice or cut into chunks.

In a large bowl, toss together the potatoes, lentils, scallion, and Baco Bits (if using). Add the mayonnaise, yogurt, salt, and pepper. Toss until completely combined.

SERVES 4

¾ pound new red potatoes (3 medium)
½ cup cooked lentils
2 tablespoons chopped scallion
1 tablespoon Baco Bits (soybean imitation-bacon pieces) (optional)
2 tablespoons mayonnaise
2 tablespoons unflavored yogurt
¼ teaspoon salt
⅛ teaspoon pepper

Calories:	183	*Total fat:*	6.0 g
Protein:	4.9 g	*Saturated fat:*	0.9 g
Carbohydrates:	28.7 g	*Cholesterol:*	4.5 mg
Fiber:	3.0 g	*Sodium:*	243 mg

❧ WHEAT BERRY, ORANGE, AND BEAN SALAD (V)

1 orange or tangerine
1½ cups cooked wheat berries (whole-grain wheat)
1¼ cups cooked or canned (drained and rinsed) cannellini
1 cup diced jicama
½ cup chopped green bell pepper
¼ cup sliced scallion
2 tablespoons chopped fresh coriander (cilantro) or parsley
2 tablespoons orange juice
1 tablespoon olive oil
1 tablespoon vegetable oil
1 tablespoon cider vinegar
1 teaspoon prepared spicy brown mustard
1 teaspoon soy sauce
¼ teaspoon salt
¼ teaspoon pepper

The textures in this salad are a perfect blend of chewy (beans and wheat berries), crunchy (jicama and green pepper), and juicy (orange). If you have problems finding jicama, use chopped pear in its place.

Using a sharp knife, remove the peel and pith from the orange. Chop the orange and set aside, reserving 2 tablespoons of the juice that has accumulated during the chopping for the dressing.

In a large bowl, combine the wheat berries, cannellini, jicama, bell pepper, scallion, and coriander.

In a small bowl, stir together the orange juice, both oils, vinegar, mustard, soy sauce, salt, and pepper. Pour over the salad and toss to combine.

Serves 6 to 8

Calories:	253	*Total fat:*	5.2 g
Protein:	13.1 g	*Saturated fat:*	0.7 g
Carbohydrates:	40.3 g	*Cholesterol:*	0
Fiber:	13.0 g	*Sodium:*	287 mg

❧ LENTIL-BULGUR SALAD WITH FETA CHEESE (E)

If you are using a mild feta cheese, you may want to add salt to this recipe. You can make this salad with only bulgur or only lentils, as long as the total is 1¾ cups.

In a large bowl, toss together the bulgur, lentils, tomato, bell pepper, onion, and parsley. Add the lemon juice, oil, oregano, and pepper and toss. Top with the feta cheese; toss again.

SERVES 4 to 6

1 cup cooled bulgur or cracked wheat
¾ cup cooked lentils
¾ cup chopped tomato
⅓ cup chopped green bell pepper
¼ cup chopped onion
¼ cup chopped parsley
1½ tablespoons fresh lemon juice
1 tablespoon olive oil
½ teaspoon dried oregano
⅛ teaspoon pepper
½ cup crumbled feta cheese

Calories:	172	Total fat:	5.1 g
Protein:	7.1 g	Saturated fat:	1.9 g
Carbohydrates:	27.3 g	Cholesterol:	8.3 mg
Fiber:	5.5 g	Sodium:	433 mg

❧ MEXICORN-LENTIL SALAD (E)(V)

Mexicorn has pieces of chopped chili and pimento in it. You can use plain corn, if that's what you have on hand, and the salad will still be delicious.

In a large bowl, combine the lentils, corn, celery, bell pepper, and scallion.

In a small bowl, stir together the oil, vinegar, garlic, paprika, and cumin. Pour over the salad and toss to combine.

SERVES 8

2 cups cooked lentils, rinsed
1 can (12 ounces) Mexicorn, drained
1 cup chopped celery
¼ cup chopped red bell pepper
¼ cup chopped scallion
3 tablespoons oil
2 tablespoons cider vinegar
1 clove garlic, minced
¼ teaspoon paprika
¼ teaspoon ground cumin

Calories:	133	*Total fat:*	5.5 g
Protein:	5.5 g	*Saturated fat:*	0.8 g
Carbohydrates:	17.2 g	*Cholesterol:*	0
Fiber:	3.3 g	*Sodium:*	282 mg

❦ MILLET AND BEAN SALAD WITH CORIANDER-LIME DRESSING (E)(Q)(V)

---

2 cups cooked millet
1 cup cooked or canned (drained and rinsed) cannellini
¼ cup chopped red onion
3 tablespoons fresh lime juice
2 tablespoons olive oil
2 tablespoons chopped fresh coriander (cilantro)
1 tablespoon grated lime rind
2 teaspoons honey
⅛ teaspoon ground red pepper

Coriander (also known as Chinese parsley or cilantro) has a very distinct and flowery flavor. Most people either adore it or absolutely hate it. If you are one of the latter, substitute chopped Italian parsley.

In a large bowl, combine the millet, cannellini, and onion.

In a small bowl, stir together the lime juice, oil, cilantro, lime rind, honey, and red pepper. Add the dressing to the salad and toss to combine thoroughly.

SERVES 4

Calories:	295	*Total fat:*	8.3 g
Protein:	11.2 g	*Saturated fat:*	1.1 g
Carbohydrates:	43.8 g	*Cholesterol:*	0
Fiber:	12.4 g	*Sodium:*	298 mg

❧ HOPPIN' JOHN SALAD (E)(V)

This recipe borrows the elements of a traditional Southern specialty. Because fatback is one of the customary ingredients (along with black-eyed peas, onion, and rice), I've included Baco Bits, which are made from soybeans, for flavor. You can omit them if you like.

In a large bowl, toss the black-eyed peas, rice, spinach, and onion until completely combined.

In a small bowl, stir together the vinegar, oil, and pepper. Pour over the salad and toss. Sprinkle on the Baco Bits and toss again.

SERVES 6 TO 8

1¼ cups cooked or canned (rinsed and drained) black-eyed peas
1½ cups cooked brown rice
1 package (10 ounces) frozen chopped spinach, thawed and well drained
⅓ cup chopped onion
3 tablespoons distilled white vinegar
3 tablespoons vegetable oil
¼ teaspoon pepper
2 tablespoons Baco Bits

Calories:	185	Total fat:	7.5 g
Protein:	6.7 g	Saturated fat:	1.1 g
Carbohydrates:	24.2 g	Cholesterol:	0
Fiber:	5.7 g	Sodium:	332 mg

❧ ONION, RICE, AND BEAN SALAD (V)

I served this dish at Thanksgiving, and all the guests commented on how much the toasted almonds added to the flavor and texture of the salad. I strongly recommend that you don't skip the toasting step.

Preheat the oven to 350° F. Bake the almonds 10 minutes, or until lightly toasted; set aside.

½ cup chopped almonds
2 tablespoons vegetable oil, divided
¾ cup chopped onion
1¼ cups water
½ cup converted rice
½ teaspoon salt, divided
1¾ cups cooked or canned (drained and rinsed) Roman (cranberry) beans

1 cup chopped tomato

¾ cup chopped celery

½ cup chopped green bell pepper

⅓ cup sliced pimento-stuffed olives

2 tablespoons red wine vinegar

2 tablespoons chopped fresh coriander (cilantro) or parsley

¼ teaspoon ground cumin

¼ teaspoon pepper

In a 1-quart saucepan, heat 1 tablespoon of the oil over medium-high heat. Add the onion and cook, stirring, until softened. Add the water and bring to a boil. Add the rice and ¼ teaspoon of the salt, and simmer, covered, 20 to 25 minutes, or until all the liquid is absorbed.

Transfer the warm rice to a large bowl and combine with the beans, tomato, celery, bell pepper, olives, and reserved almonds. Add the remaining 1 tablespoon oil, the remaining ¼ teaspoon salt, and the vinegar, coriander, cumin, and pepper. Toss until completely combined.

SERVES 10 TO 12

Calories:	172	*Total fat:*	10.4 g
Protein:	5.2 g	*Saturated fat:*	1.3 g
Carbohydrates:	18.2 g	*Cholesterol:*	0
Fiber:	5.1 g	*Sodium:*	434 mg

❧ MILLET-LENTIL SALAD WITH CINNAMON-ORANGE DRESSING   (E)(V)

---

2 cups cooked millet

1 cup cooked lentils

1 package (6 ounces) chopped dried fruit *or* ¾ cup chopped dried fruit or raisins

½ cup chopped cashews

3 tablespoons olive oil

3 tablespoons orange juice

2 tablespoons rice or cider vinegar

The packages of assorted chopped dried fruit available in the supermarket include pieces of dried apple, peach, and nectarine, along with raisins; I use Sun-Maid fruit bits for this salad. You can chop your own fruits and use any variety that you like, or use only raisins if that's more convenient. Rice vinegar usually is sold in the Oriental section of your market.

In a large bowl, combine the millet, lentils, dried fruit, and cashews.

In a small bowl, stir together the oil, orange juice, vinegar, orange rind, honey, and cinnamon. Pour over the salad and toss until combined.

SERVES 8

2 teaspoons grated orange rind
1 teaspoon honey
¼ teaspoon ground cinnamon

Calories:	219	Total fat:	9.5 g
Protein:	5.0 g	Saturated fat:	1.6 g
Carbohydrates:	31.4 g	Cholesterol:	0
Fiber:	6.7 g	Sodium:	293 mg

YUPPIE PASTA SALAD (E)(V)

The name for this dish comes from its trendiness — it's a pasta salad, and it contains fresh basil and sun-dried tomatoes. All that aside, it tastes great and is a perfect salad for a party. You can use fancy tortellini or radiatore, *but shells, ziti, or other medium-size pasta are also good.*

Cook the pasta according to package directions; drain.

In a large bowl, combine the pasta, bell pepper, zucchini, celery, beans, cherry tomatoes, olives, sun-dried tomatoes, and onion.

In a small bowl, stir together the olive oil, vinegar, lemon juice, basil, garlic, salt, and pepper. Pour the

2 cups pasta
1 cup chopped red bell pepper
1 cup sliced zucchini
1 cup sliced celery
1 cup cooked or canned (drained and rinsed) red kidney beans
12 cherry tomatoes, halved
½ cup sliced black olives
⅓ cup chopped marinated sun-dried tomatoes
⅓ cup chopped red onion
¼ cup olive oil
3 tablespoons red wine vinegar
2 tablespoons fresh lemon juice

2 tablespoons chopped
fresh basil *or* ½ teaspoon
dried basil
1 clove garlic, minced
½ teaspoon salt
¼ teaspoon pepper

dressing over the salad and toss
until combined.

SERVES 12

Calories:	161	*Total fat:*	7.4 g
Protein:	4.3 g	*Saturated fat:*	1.0 g
Carbohydrates:	21.1 g	*Cholesterol:*	0
Fiber:	3.1 g	*Sodium:*	473 mg

❧ VEGETABLE-BULGUR SALAD WITH BUTTERMILK DRESSING (E)(Q)

1 cup water
½ cup bulgur or cracked
wheat
1 cup diced cucumber
½ cup diced red bell pepper
½ cup diced zucchini
¼ cup sliced radish
¼ cup sliced scallion
3 tablespoons buttermilk
2 tablespoons mayonnaise
2 tablespoons chopped
parsley
½ teaspoon dried oregano
½ teaspoon salt
¼ teaspoon pepper

You can use reduced-calorie, fat-free, or cholesterol-free mayonnaise instead of the regular mayonnaise called for in the recipe to create a healthy, low-fat salad.

In a 1-quart saucepan, bring the water to a boil. Stir in the bulgur and return to a boil. Reduce the heat and simmer, covered, 20 minutes. Remove from the heat and let cool.

In a large bowl, combine the cooled bulgur with the cucumber, bell pepper, zucchini, radish, and scallion.

In a small bowl, using a whisk, stir together the buttermilk, mayonnaise, parsley, oregano, salt, and pepper until smooth. Pour the dressing over the salad and toss until combined.

SERVES 6

Calories:	96	Total fat:	4.0 g
Protein:	2.3 g	Saturated fat:	0.7 g
Carbohydrates:	13.4 g	Cholesterol:	3.9 mg
Fiber:	3.5 g	Sodium:	214 mg

❧ TABOULI (E)(Q)

This is a traditional salad from the Middle East. The mint is an important element, but you can reduce it to 2 tablespoons if you're not too fond of it.

Place the bulgur, tomato, cucumber, radish, parsley, scallion, and mint into a large bowl and toss to combine.

In a small bowl, stir together the yogurt, oil, vinegar, salt, and pepper. Pour over the salad and toss until combined.

SERVES 4 TO 6

1½ cups cooked bulgur or cracked wheat
¾ cup chopped tomato
¾ cup chopped cucumber
⅓ cup sliced radish
¼ cup chopped parsley
¼ cup sliced scallion
¼ cup chopped fresh mint
2 tablespoons unflavored yogurt
1 tablespoon olive oil
2 teaspoons red wine vinegar
¼ teaspoon salt
⅛ teaspoon pepper

Calories:	141	Total fat:	4.0 g
Protein:	4.7 g	Saturated fat:	0.7 g
Carbohydrates:	20.4 g	Cholesterol:	0.4 mg
Fiber:	5.3 g	Sodium:	245 mg

❧ ASIAN MILLET SALAD (E)(V)

1½ cups cooked millet
1 cup (6 ounces) diced
  tofu
½ cup chopped snow peas
½ cup frozen peas, thawed
⅓ cup chopped water chestnuts
⅓ cup chopped scallion
2 tablespoons *mirin* or dry
  sherry
1 tablespoon soy sauce
2 teaspoons vegetable oil
1½ teaspoons rice or cider
  vinegar
1 clove garlic, minced
½ teaspoon chili or sesame
  oil
½ teaspoon sugar

*Common Oriental ingredients pro-
duce a lovely combination of flavors and textures in this slightly
spicy salad. If you're not fond of
millet, use couscous instead.*

In a large bowl, combine the millet, tofu, snow peas, peas, water
chestnuts, and scallion.

In a small bowl, stir together the
mirin, soy sauce, vegetable oil, vinegar, garlic, chili oil, and sugar. Pour
over the salad and toss to combine.

SERVES 6 TO 8

Calories:	224	*Total fat:*	5.3 g
Protein:	8.2 g	*Saturated fat:*	1.3 g
Carbohydrates:	37.4 g	*Cholesterol:*	0
Fiber:	8.7 g	*Sodium:*	279 mg

❧ TOFU, QUINOA, AND HEARTS OF PALM SALAD (E)(V)

2 cups (12 ounces) diced
  tofu (¾-inch pieces)
1½ cups cooked quinoa
1 cup sliced hearts of palm
1 cup cooked fresh or frozen peas
⅓ cup sliced scallion

*This salad has only the merest hint
of curry flavor. It's a wonderful
lunch entrée on a warm summer
day.*

In a large bowl, toss together the
tofu, quinoa, hearts of palm, peas,
and scallion.

In a small bowl, mix the oils and curry powder. Stir in the lime juice, coriander, salt, and red pepper. Pour the dressing over the salad and toss to combine.

SERVES 6

2 tablespoons vegetable oil
1 tablespoon olive oil
¾ teaspoon curry powder
2 tablespoons fresh lime juice
1 tablespoon chopped fresh coriander (cilantro) or parsley
½ teaspoon salt
⅛ teaspoon ground red pepper

Calories:	165	Total fat:	8.8 g
Protein:	6.7 g	Saturated fat:	1.2 g
Carbohydrates:	16.7 g	Cholesterol:	0
Fiber:	4.2 g	Sodium:	305 mg

❧ TROPICAL SALAD WITH CALYPSO DRESSING (E)(Q)(V)

I originally made this salad with a combination of peaches and papaya, but I found the flavors of the two fruits so similar that I opted for only the papaya. If you are having trouble finding papaya, feel free to substitute fresh peaches or nectarines.

In a large bowl, combine the couscous, bean sprouts, papaya, cucumber, and kiwi.

In a small bowl, stir together the yogurt, lime juice, oil, coriander, scallion, curry powder, ginger, salt, and red pepper. Add the dressing to the salad and toss until combined.

SERVES 4 to 6

1½ cups cooked couscous
1 cup bean sprouts
1 cup chopped papaya
½ cup chopped cucumber
⅓ cup chopped kiwi
¼ cup unflavored yogurt
3 tablespoons fresh lime juice
1 tablespoon olive oil
2 tablespoons chopped fresh coriander (cilantro) or parsley
1 tablespoon chopped scallion
½ teaspoon curry powder
½ teaspoon grated fresh ginger

½ teaspoon salt
⅛ teaspoon ground red
    pepper

Calories:	179	*Total fat:*	5.1 g
Protein:	5.4 g	*Saturated fat:*	1.2 g
Carbohydrates:	33.7 g	*Cholesterol:*	1.0 mg
Fiber:	4.9 g	*Sodium:*	460 mg

❧ BEAN SPROUT AND WATERCRESS SALAD WITH GRAPEFRUIT-MINT DRESSING (E)(Q)(V)

1 grapefruit
2 cups mung bean sprouts
2 cups watercress
½ cup sliced endive
¼ cup sliced red onion
2 tablespoons raisins
1 tablespoon sunflower
    seeds
2 sprigs fresh mint
2 tablespoons olive oil
2 tablespoons grapefruit
    juice
1½ teaspoons distilled white
    vinegar
½ teaspoon dry mustard
½ teaspoon sugar
⅛ teaspoon salt

This salad has a fresh, unusual flavor. Serve it as soon as you add the dressing so the greens don't become soggy. It makes an excellent appetizer or side dish.

Using a sharp paring knife, remove the peel and white pith from the grapefruit. Remove the segments by slicing between the tough membrane and the flesh. Squeeze enough segments as necessary to get 2 tablespoons juice; set aside for the dressing. Place the remaining segments in a large bowl.

Add the sprouts, watercress, endive, onion, raisins, and sunflower seeds to the bowl with the grapefruit segments.

Put the mint, oil, juice, vinegar, mustard, sugar, and salt into a blender container. Cover and blend until the mint is finely chopped. Pour the dressing over the salad and toss.

SERVES 4

Calories:	134	Total fat:	8.1 g
Protein:	3.4 g	Saturated fat:	1.2 g
Carbohydrates:	15.3 g	Cholesterol:	0
Fiber:	3.2 g	Sodium:	87 mg

❧ GREEN BEAN AND ENDIVE SALAD WITH WALNUT-LEEK DRESSING (E)(V)

The sautéed leeks add a mild oniony flavor to this dressing. To make a special salad, toss in a handful of fresh raspberries.

Heat the oil in a small skillet over medium heat. Add the leek and cook, stirring, until softened. Remove from the heat and stir in the vinegar, mustard, salt, and pepper; set aside.

Cook the green beans in boiling water until tender-crisp. Drain and rinse under cold water to cool. Place the green beans in a large bowl; add the endive, apple, raisins, and walnuts. Add the dressing and toss until combined.

SERVES 6

1½ tablespoons olive oil
½ cup sliced leek (white and light green parts only), thoroughly rinsed
1 tablespoon balsamic vinegar
1 teaspoon Dijon mustard
½ teaspoon salt
⅛ teaspoon pepper
½ pound whole green beans, ends trimmed
2 heads endive, cut into julienne strips
¾ cup chopped apple
½ cup raisins
½ cup walnut pieces

Calories:	156	Total fat:	9.8 g
Protein:	2.7 g	Saturated fat:	1.0 g
Carbohydrates:	17.5 g	Cholesterol:	0
Fiber:	3.3 g	Sodium:	190 mg

❧ CHUNKY SALAD WITH GAZPACHO DRESSING (E)(Q)(V)

2 cups peeled, seeded, and cubed cucumber
2 cups sliced celery (1-inch pieces)
1 cup cubed green bell pepper (¾-inch pieces)
½ cup sliced scallion (½-inch pieces)
¼ cup V8 juice
2 tablespoons tomato paste
1½ tablespoons vegetable oil
1 tablespoon red wine vinegar
½ teaspoon salt
¼ teaspoon Worcestershire sauce
¼ teaspoon dried basil
¼ teaspoon celery seed
¼ teaspoon Red Hot sauce

Gazpacho is a cold soup made of pureed vegetables and oil (see page 89 for a recipe). This salad incorporates all the traditional gazpacho ingredients. Feel free to choose plain tomato juice, spicy V8, low-sodium V8, or tomato juice instead of the V8 called for in the recipe.

In a large bowl, toss together the cucumber, celery, bell pepper, and scallion.

In a small bowl, mix the juice, tomato paste, oil, vinegar, salt, Worcestershire sauce, basil, celery seed, and Red Hot sauce. Pour over the salad and toss. Let stand 1 hour before serving.

SERVES 4 to 6

Calories:	83	*Total fat:*	5.5 g
Protein:	1.6 g	*Saturated fat:*	0.8 g
Carbohydrates:	8.7 g	*Cholesterol:*	0
Fiber:	2.9 g	*Sodium:*	383 mg

❧ DICED VEGETABLE SALAD (E)(Q)(V)

¾ cup diced cucumber
½ cup diced green bell pepper
½ cup diced red bell pepper
½ cup diced yellow or orange bell pepper
½ cup diced radish

All the vegetables in this salad should be diced into ⅛-inch pieces. It can be served as a relish or chutney, as well as a salad. You can make this dish with only one or two of the different bell peppers — just use a total of 1½ cups — although it is much prettier when you have all three colors.

In a medium bowl, stir together the cucumber, all the bell pepper, and the radish, onion, vinegar, oil, salt, and pepper.

SERVES 4

¼ cup diced red onion
1 tablespoon red wine vinegar
1 teaspoon olive oil
½ teaspoon salt
¼ teaspoon pepper

Calories:	29	Total fat:	1.5 g
Protein:	0.7 g	Saturated fat:	0.2 g
Carbohydrates:	4.4 g	Cholesterol:	0
Fiber:	1.5 g	Sodium:	224 mg

❧ NAKED TOMATO SALAD (E)(Q)(V)

"Naked" because it has no dressing, this is my very favorite "diet" salad. It has no fat and hardly any calories. My neighbor Marie Riesel says her mother makes the same salad but sprinkles it with oregano. Sounds good.

Toss all the ingredients together in a medium bowl. Let stand at least 10 minutes before serving.

SERVES 4

3 cups chopped fresh, ripe tomatoes (¾-inch pieces)
⅔ cup chopped onion (¼-inch pieces)
½ teaspoon salt
½ teaspoon pepper

Calories:	35	Total fat:	0
Protein:	1.8 g	Saturated fat:	0
Carbohydrates:	7.9 g	Cholesterol:	0
Fiber:	1.6 g	Sodium:	278 mg

❧ CURRIED BROCCOFLOWER SALAD (E)(Q)

3 cups broccoflower florets
1 cup sliced Jerusalem artichoke or jicama
½ cup julienned red bell
  pepper
12 pitted black olives,
  halved
¼ cup sliced scallion
2 tablespoons white or red
  wine vinegar
2 tablespoons buttermilk
1 tablespoon vegetable oil
½ teaspoon curry powder
¼ teaspoon salt
⅛ teaspoon ground cumin
⅛ teaspoon ground red
  pepper

Broccoflower is a cross between broccoli and cauliflower — it looks like a slightly pointy, bright green cauliflower. If you can't find broccoflower, use cauliflower instead.

Add the broccoflower to boiling water and cook 1 to 2 minutes, or until slightly cooked. Drain and cool.

In a large bowl, toss together the drained broccoflower and the Jerusalem artichoke, bell pepper, olives, and scallion.

In a small bowl, mix the vinegar, buttermilk, oil, curry powder, salt, cumin, and red pepper until completely combined. Pour the dressing over the salad and toss.

SERVES 6 TO 8

Calories:	67	Total fat:	3.0 g
Protein:	2.9 g	Saturated fat:	0.8 g
Carbohydrates:	10.4 g	Cholesterol:	0
Fiber:	3.1 g	Sodium:	184 mg

❧ PEA SALAD (E)(Q)

2 cups cooked fresh or frozen peas, chilled
¾ cup diced red bell pepper
½ cup thinly sliced celery
¼ cup sliced scallion

This salad is easy to prepare and goes well with almost anything. When I use frozen peas, I don't bother cooking them — I just defrost them.

In a medium bowl, combine the peas, bell pepper, celery, and scallion.

In a small bowl, stir together the yogurt, mayonnaise, dill, mustard, salt, Worcestershire sauce, and pepper. Pour over the peas and toss. Cover and chill.

SERVES 4 to 6

¼ cup unflavored yogurt
1 tablespoon mayonnaise
1 tablespoon chopped fresh dill
1 teaspoon Dijon mustard
¼ teaspoon salt
⅛ teaspoon Worcestershire sauce
⅛ teaspoon pepper

Calories:	108	Total fat:	3.4 g
Protein:	5.5 g	Saturated fat:	0.6 g
Carbohydrates:	14.9 g	Cholesterol:	2.1 mg
Fiber:	4.2 g	Sodium:	265 mg

GREEN BEAN–KOHLRABI SALAD (E)(Q)

If you can't find kohlrabi in the market, you can use rutabaga, daikon radish, peeled and sliced broccoli stalk, or any other vegetable that's crunchy.

In a large bowl, toss together the green beans, kohlrabi, and tomatoes.

In a small bowl, stir together the mayonnaise, lemon juice, honey mustard, poppy seeds, celery seed, salt, and pepper. Add the dressing to the salad and toss to combine completely.

SERVES 4

2 cups cooked fresh or frozen cut green beans
½ cup peeled, thinly sliced kohlrabi
6 cherry tomatoes, quartered
1½ tablespoons mayonnaise
1½ tablespoons fresh lemon juice
2 teaspoons honey mustard
½ teaspoon poppy seeds
¼ teaspoon celery seed
¼ teaspoon salt
⅛ teaspoon pepper

Calories:	91	Total fat:	6.2 g
Protein:	2.2 g	Saturated fat:	0.9 g
Carbohydrates:	9.7 g	Cholesterol:	4.3 mg
Fiber:	4.1 g	Sodium:	221 mg

❦ TOSSED SALAD WITH AVOCADO AND BLUE CHEESE DRESSING (E)(Q)

⅓ cup buttermilk
3 tablespoons mayonnaise
2 tablespoons blue cheese
⅛ teaspoon pepper
    Dash Worcestershire
    sauce
4 cups romaine pieces
1 medium tomato,
    chopped
½ avocado, diced
½ chopped red bell pepper
½ small red onion, sliced

Of course you can use any combination of vegetables in this salad, but the avocado adds a nice texture and richness. Replace all or part of the mayonnaise with fat-free mayonnaise for a lower-calorie version of the dressing.

Place the buttermilk, mayonnaise, blue cheese, pepper, and Worcestershire sauce in a blender container. Cover and blend until well combined.

In a large bowl, toss together the lettuce, tomato, avocado, bell pepper, and onion. Serve with the dressing on the side or poured over the salad and tossed.

SERVES 4

Calories:	170	_Total fat:_	14.5 g
Protein:	4.2 g	_Saturated fat:_	3.3 g
Carbohydrates:	7.8 g	_Cholesterol:_	12.1 mg
Fiber:	4.2 g	_Sodium:_	189 mg

❦ WILTED CUCUMBER SALAD (E)(V)

5 cups peeled, thinly sliced
    cucumber
¾ cup thinly sliced onion
1½ tablespoons salt
1 cup cold water
2 teaspoons distilled white
    vinegar

This is an old family recipe, which we serve on the most festive occasions. The best part is that there's no oil in the dressing. However, if you're watching your sodium intake, this recipe is not for you!

Layer the cucumber and onion in a large bowl, sprinkling each layer liberally with the salt. Place a plate almost the diameter of the bowl on top of the vegetables and put a weighted object on top of the plate (I normally use a quart jar filled with water). Let stand 30 minutes.

Drain all of the liquid from the bowl and then, picking up small handfuls of the salted vegetables, squeeze them between your hands until no more liquid drips out. Place that small handful into a clean bowl and continue with the remaining vegetables. Taste the cucumbers; if they are still so salty that they make you squint, rinse them under cold water until they are to your taste.

In a small bowl, stir together the water, vinegar, sugar, and pepper until the sugar dissolves. Pour over the vegetables and toss until combined. Chill before serving.

1½ teaspoons sugar
⅛ teaspoon pepper

SERVES 6

Calories:	25	*Total fat:*	0.2 g
Protein:	0.8 g	*Saturated fat:*	0
Carbohydrates:	5.8 g	*Cholesterol:*	0
Fiber:	1.5 g	*Sodium:*	440 mg

∾ GUILT-FREE ROQUEFORT DRESSING (E)(Q)

3 tablespoons buttermilk
1½ teaspoons Roquefort
1 tablespoon chopped
onion
Pinch pepper
Dash Worcestershire
sauce
¼ cup unflavored yogurt

You can make this and the next three dressings as guilt-free as you like by using low-fat or nonfat ingredients instead of the full-fat ones.

Place the buttermilk, Roquefort, onion, pepper, and Worcestershire sauce in a blender container. Cover and blend until smooth. Fold into the yogurt.

MAKES ABOUT ½ CUP

PER TABLESPOON USING FULL-FAT PRODUCTS:

Calories:	10	*Total fat:*	0.6 g
Protein:	0.6 g	*Saturated fat:*	0.4 g
Carbohydrates:	0.7 g	*Cholesterol:*	2.0 mg
Fiber:	0	*Sodium:*	26 mg

PER TABLESPOON USING NONFAT PRODUCTS:

Calories:	10	*Total fat:*	0.4 g
Protein:	0.8 g	*Saturated fat:*	0.2 g
Carbohydrates:	0.9 g	*Cholesterol:*	1.1 mg
Fiber:	0	*Sodium:*	28 mg

✎ GUILT-FREE THOUSAND ISLAND DRESSING (E)(Q)

In a small bowl, stir together the yogurt, mayonnaise, parsley, catsup, and relish.

MAKES ABOUT ½ CUP

¼ cup unflavored yogurt
2 tablespoons mayonnaise
2 tablespoons chopped parsley
1 tablespoon catsup
2 teaspoons India or pickle relish

PER TABLESPOON USING FULL-FAT PRODUCTS:

Calories:	33	*Total fat:*	3.0 g
Protein:	0.4 g	*Saturated fat:*	0.6 g
Carbohydrates:	1.5 g	*Cholesterol:*	3.0 mg
Fiber:	0.1 g	*Sodium:*	54.3 mg

PER TABLESPOON USING NONFAT PRODUCTS:

Calories:	8	*Total fat:*	0
Protein:	0.5 g	*Saturated fat:*	0
Carbohydrates:	1.7 g	*Cholesterol:*	0
Fiber:	0.1 g	*Sodium:*	56 mg

✎ GUILT-FREE HERB DRESSING (E)(Q)

Place ¼ cup of the yogurt, along with the scallion, parsley, mayonnaise, garlic, capers, and basil, in a blender container. Cover and blend until the herbs are minced. Pour into a small bowl. Fold in the remaining ¼ cup yogurt.

MAKES ABOUT ⅔ CUP

½ cup unflavored yogurt, divided
¼ cup sliced scallion
2 tablespoons chopped parsley
1 tablespoon mayonnaise
1 clove garlic, minced
1 teaspoon capers
¼ teaspoon dried basil

PER TABLESPOON USING FULL-FAT PRODUCTS:

Calories:	18	*Total fat:*	1.5 g
Protein:	0.5 g	*Saturated fat:*	0.4 g
Carbohydrates:	0.8 g	*Cholesterol:*	2.3 mg
Fiber:	0	*Sodium:*	13.5 mg

PER TABLESPOON USING NONFAT PRODUCTS:

Calories:	7	*Total fat:*	0
Protein:	0.7 g	*Saturated fat:*	0
Carbohydrates:	1.7 g	*Cholesterol:*	0
Fiber:	0	*Sodium:*	16.8 mg

❧ GUILT-FREE GARLIC-PEPPER DRESSING (E)(Q)

¼ cup unflavored yogurt
¼ cup buttermilk
1 tablespoon mayonnaise
2 cloves garlic, minced
¼ teaspoon coarsely
   ground pepper
⅛ teaspoon seasoned salt

Stir together the yogurt, buttermilk, mayonnaise, garlic, pepper, and seasoned salt in a small bowl.

MAKES ABOUT ½ CUP

PER TABLESPOON USING FULL-FAT PRODUCTS:

Calories:	20	*Total fat:*	1.7 g
Protein:	0.5 g	*Saturated fat:*	0.4 g
Carbohydrates:	0.8 g	*Cholesterol:*	2.2 mg
Fiber:	0	*Sodium:*	55 mg

PER TABLESPOON USING NONFAT PRODUCTS:

Calories:	8	*Total fat:*	0.1 g
Protein:	0.7 g	*Saturated fat:*	0
Carbohydrates:	1.0 g	*Cholesterol:*	0
Fiber:	0	*Sodium:*	55 mg

ぐ TAHINI DRESSING (E)(Q)(V)

Tahini is a popular ingredient in Middle Eastern cooking. This dressing is usually served on falafel. You can also use it as a dip or spread, served with crudités or pita wedges.

Stir together the tahini, lemon juice, water, parsley, and garlic in a small bowl.

MAKES ABOUT ½ CUP

¼ cup tahini (sesame paste)
2 tablespoons fresh lemon juice
2 tablespoons water
1 tablespoon chopped parsley
1 small clove garlic, minced

Calories:	93	*Total fat:*	8.5 g
Protein:	2.7 g	*Saturated fat:*	1.2 g
Carbohydrates:	3.4 g	*Cholesterol:*	0
Fiber:	1.6 g	*Sodium:*	6 mg

Breads and Spreads

❧

BAKING WITH YEAST

THERE ARE AS MANY VARIABLES when baking with yeast as there are when cooking grains and beans. The following are a few of the things I've learned about bread baking that I hope will help you as you use the recipes.

Activating the yeast in water (or other liquid) that is the correct temperature is the first and most important step in working with yeast. The ideal temperature range is 105° to 115° F, and the best way to determine the water temperature is with an instant-read thermometer. If you don't have an instant-read thermometer, test the temperature by touch; it should be very warm, but not hot. Since yeast is a living organism, water that is too hot will kill the yeast, and water that is too cold may not activate it at all. In either case your bread will never rise.

Proofing is the process of checking whether the yeast is active. Your recipe will direct you to stir the yeast into a moderate amount of water and let it stand until it starts to form a layer of foam (adding a little sugar will speed up the foaming). This foam indicates that the yeast has been activated. If the foam has not appeared within 10 to 15 minutes (with sugar the mixture will foam in about 3 minutes), discard the mixture and try again; your water may have been too hot or too cold.

One other reason foaming may not occur is that the yeast is too old. There is an expiration date on most packages of yeast. Check to make sure the date has not passed. You can extend the life of your yeast somewhat by storing it in the refrigerator, rather than at room temperature.

Mixing the dough follows a successful proofing of the yeast: the wet ingredients are stirred together with most of the flour(s) to form a dough that is manageable. This means that the dough is dry enough that you can form a soft ball with it, and that you do not have to scrape it off your hands with a knife after you touch it. It may, however, still be somewhat sticky. When I finish mixing the dough, I put about 1/3 cup of flour on the work surface. I turn the dough out onto the flour and turn the ball of dough over so that it is completely coated with flour. Then I start kneading.

Kneading the dough is the building up of gluten to make the dough elastic and stretchy, allowing it to rise as the yeast forms carbon dioxide air bubbles. To knead, fold the dough ball in half. Then, using the heel of your hand, press downward and away from you at the same time. Turn the dough a quarter turn, fold in half, and press down and away again. When the dough starts to stick to your hands, add more flour to the work surface. Continue this folding and pressing process for about 7 minutes. It's almost impossible to overknead bread dough.

Place the dough in a large greased bowl, and rotate the dough so that it is coated with grease. Cover with plastic wrap and place in a warm, draft-free spot to *double in bulk*. You can determine whether the dough has doubled in two ways. Before setting the dough aside to rise, you can mark the bowl with a grease pencil at the point you think is twice as high as the kneaded, unrisen dough. The second way is to wait until the dough has visibly puffed up and then press it lightly, using your fingertip. A slight indentation should have formed. If the dough is doubled in bulk, the indentation will remain where you pressed it. If it begins to disappear, the dough can rise some more. Doubling in bulk usually takes about 1 to 1½ hours but will vary, depending on the temperature of the room, the temperature of the water when the yeast was proofed, how much salt is in the dough, whether there are acid ingredients, and what type of flour is being used. As long as your yeast foamed, your bread should rise; it just may take longer than 1½ hours.

Punching the dough down is deflating the dough by pressing it down with your fist. At this point you can shape the dough into whatever size or shape the recipe calls for.

Second rising and baking. Place the dough on a greased baking sheet or into a greased pan, cover with greased plastic wrap, and let rise again. This second rising will take about half the time of the first rising. Be aware that it is possible to ruin your bread by letting it overrise. The yeast will eventually collapse, and that will be that. So don't just go out and forget about the dough you've left to rise (this is true for both risings).

You can make decorative slashes in your loaves either before the second rising (once the dough is shaped) or after the second rising.

The final step is *baking the bread* the specified amount of time, or until the crust is golden-brown and sounds hollow when you tap it. Remove the loaf from the pan or baking sheet and let cool on a wire rack.

Finishing touches. For a crispier crust, bake the bread in a moist atmosphere. You can create moisture by placing a pan of water in the bottom of the oven while the bread bakes. Other methods are tossing a few ice cubes into the oven when you start baking (not advisable with an electric oven) and spraying water into the oven.

For a professional-looking bread, brush your loaves with beaten egg white before baking. To create a less shiny finish, brush with beaten whole eggs. Brushing the loaves with egg yolk mixed with a little water will produce a dark matte finish.

❧ MOLASSES-OATMEAL BREAD

1½ cups boiling water
2 cups old-fashioned rolled oats, divided
½ cup very warm water (105°–115° F)
½ teaspoon sugar
2 packages active dry yeast
1 cup milk
½ cup molasses
1 tablespoon salt
1 egg
3 cups whole wheat flour
½ cup oat bran
2 to 3 cups all-purpose flour

I find this dense, moist bread exceptionally good for toasting. For a prettier loaf, sprinkle extra oats on top before the second rising.

Grease two 8½ × 4½ × 2¾-inch loaf pans; set aside.

In a large bowl, stir together the boiling water and 1½ cups of the oats. Let stand until the water is absorbed.

In a glass measuring cup, stir together the warm water and sugar. Stir in the yeast and let stand until ¼ inch of bubbly foam forms on top.

Add the milk, molasses, and salt to the oat and water mixture; stir in the egg. Stir in the yeast mixture. Add the whole wheat flour, the oat bran, and the remaining ½ cup oats. Stir in 1½ cups of the all-purpose flour to make a dough that is easy to handle.

Turn the dough out onto a well-floured surface and knead in enough of the remaining flour to make a dough that is smooth, elastic, and no longer sticky. Place the dough in a large greased bowl, turn to coat all over with the grease, and cover with greased plastic wrap. Set in a warm, draft-free spot until doubled in bulk.

Punch the dough down and form into 2 loaves. Place in the prepared pans. Cover with greased plastic wrap and let rise until doubled in bulk.

Preheat the oven to 350° F. Bake for 50 to 60 minutes, or until the loaves are browned on top and bottom. Remove from the pans and cool on a wire rack.

MAKES 2 LOAVES (14 SLICES EACH)

Calories:	142	*Total fat:*	1.5 g
Protein:	5.3 g	*Saturated fat:*	0.4 g
Carbohydrates:	28.4 g	*Cholesterol:*	8.6 mg
Fiber	3.3 g	*Sodium:*	242 mg

☙ GRAHAM BREAD (V)

This is a very coarse-grained bread with lots of crust. The hearty flavor stands up to any stew or other strongly flavored dish.

In a glass measuring cup, stir together the warm water and sugar. Stir in the yeast and let stand until ¼ inch of bubbly foam forms on top.

½ cup very warm water
  (105°–115° F)
½ teaspoon sugar
 2 packages active dry yeast
 3 cups graham flour
1¾ to 2½ cups all-purpose
  flour
 2 teaspoons salt
1½ cups water

3 tablespoons pure maple
   syrup
2 tablespoons vegetable oil

In a large bowl, combine 2 cups of the graham flour, 1 cup all-purpose flour, and the salt. Stir in the yeast mixture and the water, maple syrup, and oil. Stir in the remaining 1 cup graham flour and as much of the remaining all-purpose flour as necessary to make a dough that is easy to handle.

Turn the dough out onto a well-floured surface and knead in enough of the remaining all-purpose flour to make a dough that is smooth, elastic, and no longer sticky. Place the dough in a large greased bowl, turn to coat all over with the grease, and cover with greased plastic wrap. Set in a warm, draft-free spot until doubled in bulk.

Punch the dough down and form into 2 round loaves. Place on greased baking sheets. Cover with greased plastic wrap and let rise until doubled in bulk.

Preheat the oven to 350° F. Bake for 50 minutes, or until the loaves are browned on top and bottom and sound hollow when tapped. Cool on a wire rack.

MAKES 2 LOAVES (12 SLICES EACH)

Calories:	72	*Total fat:*	1.5 g
Protein:	2.7 g	*Saturated fat:*	0.1 g
Carbohydrates:	12.4 g	*Cholesterol:*	0
Fiber:	1.6 g	*Sodium:*	180 mg

❧ ANADAMA BREAD (V)

Anadama bread has a faint grittiness from the cornmeal and a distinct yeasty flavor. To ensure that this free-form loaf rises high instead of just spreading, be careful to knead in enough flour to make the dough quite stiff.

In a glass measuring cup, stir together the warm water and sugar. Stir in the yeast and let stand until ¼ inch of bubbly foam forms on top.

In a large bowl, combine 2 cups of the flour with the cornmeal and salt. Stir in the yeast mixture and the water, molasses, and oil. Stir in between ¼ and 1 cup more flour to make a dough that is easy to handle.

Turn the dough out onto a well-floured surface and knead in enough of the remaining flour to make a dough that is stiff and no longer sticky. Place the dough in a large greased bowl, turn to coat all over with the grease, and cover with greased plastic wrap. Set in a warm, draft-free spot until doubled in bulk.

Punch the dough down and form into a 6-inch-round loaf. Place on a greased baking sheet. Cover with greased plastic wrap and let rise until doubled in bulk.

Preheat the oven to 350° F. Bake for 50 minutes, or until the loaf is browned on top and bottom and sounds hollow when tapped. Cool on a wire rack.

MAKES 1 ROUND LOAF (18 SLICES)

½ cup very warm water (105°–115° F)
½ teaspoon sugar
1 package active dry yeast
3½ to 4 cups all-purpose flour
¾ cup white or yellow cornmeal
2 teaspoons salt
¾ cup water
¼ cup molasses
3 tablespoons vegetable oil

Calories:	146	*Total fat:*	2.2 g
Protein:	3.7 g	*Saturated fat:*	0.3 g
Carbohydrates:	26.9 g	*Cholesterol:*	0
Fiber:	1.3 g	*Sodium:*	240 mg

✂ LIGHT RYE BREAD (V)

3 cups very warm water (105°–115° F)

3 cups rye flour, divided

2 packages active dry yeast, divided

3 tablespoons cane syrup or honey

2 tablespoons caraway seeds (optional)

2 tablespoons vegetable oil

1 tablespoon salt

4 to 4½ cups all-purpose flour, divided

This recipe makes two large loaves, but as long as you're going to the bother of making a starter, you might as well have some extra bread for your freezer when you're done. The finished bread is light in texture, although not in flavor or color. If you don't want round loaves, you can bake this bread in two 9 × 5 × 3-inch loaf pans, but be sure to grease the pans well.

In a large bowl, stir together the water, 2 cups of the rye flour, and 1 package of the yeast. Cover with aluminum foil and let stand at room temperature 24 hours.

Stir the fermented yeast mixture, mixing in any crust that may have formed on top. Add the remaining 1 package yeast to the fermented mixture and let stand 10 minutes. Stir in the syrup, caraway seeds (if using), oil, and salt until combined. Stir in the remaining 1 cup rye flour and 3½ to 4 cups of the all-purpose flour until the dough is soft but manageable.

Turn the dough out onto a heavily floured surface and knead about 7 minutes, using as much of the remaining all-purpose flour as necessary to create a dough that is smooth and elastic (it may still be slightly tacky).

Place the dough in a very large greased bowl (I use a 6-quart pot for this), turning the dough once so the top is oiled. Cover with plastic wrap and let rise in a warm, draft-free spot until doubled in bulk (about 1¼ hours). Punch the dough down and divide in half. Form into 2 round loaves and place each on a greased baking sheet. Using a sharp knife, cut 3 slashes into the top of each loaf. Cover with greased plastic wrap and let rise until doubled in bulk (about 30 minutes).

Preheat the oven to 400° F and bake 10 minutes. (For a very crispy crust, toss 2 ice cubes onto the oven floor and close the oven door to allow steam to form in the oven. I hesitate to do this in an electric oven; if you have an electric oven, spray with water instead.) Reduce the heat to 350° and bake 40 minutes longer, or until the loaves are browned on top and bottom and sound hollow when tapped. Remove from the baking sheets and cool on wire racks.

MAKES 2 LARGE LOAVES (ABOUT 20 SLICES EACH)

Calories:	90	Total fat:	1.0 g
Protein:	2.5 g	Saturated fat:	0.1 g
Carbohydrates:	18.0 g	Cholesterol:	0
Fiber:	1.2 g	Sodium:	160 mg

❧ CORN RYE (V)

3 cups cold water, divided
½ cup white or yellow
   cornmeal
2 tablespoons oil
2 cups rye flour
4 to 5 cups graham or
   whole wheat flour
1 tablespoon salt
3 tablespoons barley malt
   or molasses
½ cup very warm water
   (105°–115° F)
½ teaspoon sugar
2 packages active dry yeast

I find this rye bread a very satisfying loaf, especially when toasted. It is dense, with the characteristic sourness of rye. If you can't find barley malt, you can substitute molasses.

In a 1-quart saucepan, stir together 2 cups of the cold water and the cornmeal. Bring to a boil and remove from the heat. Stir in the remaining 1 cup cold water and the oil. Cool 20 minutes.

In a large bowl, stir together the rye flour, 2 cups of the graham flour, and the salt, cornmeal mixture, and barley malt.

In a glass measuring cup, mix the warm water and sugar. Stir in the yeast and let stand until ¼ inch of bubbly foam forms on top. Add the proofed yeast to the cornmeal-flour mixture. Stir in 1½ cups more graham flour to form a soft dough. Cover with a damp towel and let rise 1 hour.

Punch down the dough and stir in ½ cup more graham flour. Turn out onto a well-floured board and knead in as much of the remaining flour as necessary to form a dough that is smooth and elastic. Place in a greased bowl, turning the dough to grease the top. Cover with greased plastic wrap and let rise in a warm, draft-free place until doubled in bulk.

Punch down the dough and form into 2 loaves. Place each in a greased

9 × 5 × 2¾-inch loaf pan. Cover with greased plastic wrap and let rise in a warm, draft-free place until doubled in bulk.

Preheat the oven to 375° F. Bake 1 hour, or until browned on top and bottom.

MAKES 2 LOAVES (14 SLICES EACH)

Calories:	112	*Total fat:*	1.7 g
Protein:	3.5 g	*Saturated fat:*	0.5 g
Carbohydrates:	21.0 g	*Cholesterol:*	0
Fiber:	3.0 g	*Sodium:*	227 mg

❧ THREE-GRAIN BREAD (V)

You'll be helping yourself to slice after slice of this very moist bread, until it's gone before you know it.

In a glass measuring cup, stir together the warm water and sugar. Stir in the yeast and let stand until ¼ inch of bubbly foam forms on top.

In a large bowl, combine the whole wheat flour, rye flour, cornmeal, and salt. Stir in the yeast mixture, water, oil, and honey. Stir in 1 cup of the all-purpose flour to make a dough that is easy to handle.

Turn the dough out onto a well-floured surface and knead in enough of the remaining flour to make a dough that is smooth, elastic, and no longer sticky. Place the dough in a large greased bowl, turn to coat all over with grease, and cover with greased plastic wrap. Set

½ cup very warm water (105°–115° F)
½ teaspoon sugar
1 package active dry yeast
1 cup whole wheat flour
1 cup rye flour
½ cup yellow cornmeal
2 teaspoons salt
¾ cup water (or milk)
3 tablespoons vegetable oil
3 tablespoons honey
1¼ to 1¾ cups all-purpose flour

in a warm, draft-free spot until doubled in bulk.

Punch down and shape into a round loaf. Place on a greased baking sheet. Cover with greased plastic wrap and let rise until doubled in bulk.

Preheat the oven to 350° F. Bake for 40 to 50 minutes, or until the loaf is browned on top and bottom and sounds hollow when tapped. Cool on a wire rack.

MAKES 1 LOAF (18 SLICES)

Calories:	134	*Total fat:*	2.7 g
Protein:	3.4 g	*Saturated fat:*	0.4 g
Carbohydrates:	24.8 g	*Cholesterol:*	0
Fiber:	2.0 g	*Sodium:*	238 mg

❧ PUMPERNICKEL (V)

2 cups water, divided
⅓ cup white or yellow cornmeal
⅓ cup molasses
3 tablespoons unsweetened cocoa
1 tablespoon instant coffee
3 tablespoons firmly packed light or dark brown sugar
1½ tablespoons caraway seeds
1½ teaspoons fennel seeds (optional)
1 tablespoon salt
½ cup very warm water (105°–115° F)
½ teaspoon sugar
2 packages active dry yeast

This bread is a cross between a real pumpernickel bread, which is like a very heavy rye, and Russian black bread, which has cocoa and coffee flavors. You can add raisins for a special taste and texture.

In a medium saucepan, stir together 1 cup of the water and the cornmeal. Bring to a boil and cook, stirring frequently, until the mixture is the consistency of cereal. Whisk in the molasses, cocoa, and coffee. Stir in the remaining 1 cup water and the brown sugar, caraway and fennel seeds, and salt.

In a glass measuring cup, mix the warm water and sugar. Stir in the yeast and let stand until ¼ inch of bubbly foam forms on top.

While the yeast is proofing, stir together the rye and graham flours in a large bowl. Add the cornmeal and yeast mixtures. Stir in 1¾ cups of the bread flour to form a manageable dough.

Turn the dough out onto a heavily floured surface and knead about 7 minutes, using as much of the remaining bread flour as necessary to create a dough that is smooth and elastic. (It may still be slightly tacky.)

Place the dough in a very large greased bowl (I use a 6-quart pot), turning the dough once so the top is oiled. Cover with greased plastic wrap and let rise in a warm, draft-free spot until doubled in bulk (about 1¼ hours).

Punch the dough down and divide in half. Form into 2 round loaves and place each on a greased baking sheet. Using a sharp knife, cut 3 slashes into the top of each loaf. Cover with greased plastic wrap and let rise until doubled in bulk.

Preheat the oven to 350° F. Bake 50 minutes, or until the loaves are browned on top and bottom and sound hollow when tapped. Remove from the baking sheets and cool on a wire rack.

MAKES 2 LARGE LOAVES (ABOUT 20 SLICES EACH)

2 cups rye flour
1 cup graham or whole wheat flour
2¼ to 2¾ cups bread flour, divided

Calories:	82	Total fat:	0.4 g
Protein:	3.2 g	Saturated fat:	0
Carbohydrates:	16.2 g	Cholesterol:	0
Fiber:	2.2 g	Sodium:	163 mg

❧ WHOLE WHEAT BAGUETTE (V)

½ cup very warm water
  (105°–115° F)
1 teaspoon sugar
1 package active dry yeast
1 cup whole wheat flour
2 to 2½ cups bread flour,
  divided
1½ teaspoons salt
¾ cup water
1 egg white, beaten (op-
  tional)
6 ice cubes

I call for bread flour in this recipe because French bread is best when made with a high-gluten flour. If you can't find bread flour, all-purpose flour will do.

In a glass measuring cup, stir together the warm water and sugar. Stir in the yeast and let stand until ¼ inch of bubbly foam forms on top.

While the yeast is proofing, stir together the whole wheat flour, ½ cup of the bread flour, and the salt in a large bowl. Add the yeast mixture and water; stir until combined. Stir in ½ cup of the bread flour. The dough should be slightly sticky but should not cling to your hands.

Turn the dough out onto a heavily floured surface and knead about 7 minutes, using as much of the remaining bread flour as necessary to create a dough that is smooth and elastic (it may still be slightly tacky).

Place the dough in a greased bowl, turning once so the top is oiled. Cover with greased plastic wrap and let rise in a warm, draft-free spot until doubled in bulk.

Punch the dough down and divide into 4 pieces. Roll each piece into a 16-inch log. Place 2 logs on each of 2 greased baking sheets. Brush each loaf with the egg white (if using) and then, with a sharp knife, cut diagonal slashes in the top of each loaf. Let stand, covered with greased waxed paper, 30 minutes.

Preheat the oven to 425° F and place the loaves in the oven. (For a crisper crust, toss 2 ice cubes into the oven and close the door immediately. When the ice cubes have melted, toss in another 2 cubes. Repeat with the remaining 2 ice cubes. If you have an electric oven, spray with water instead.) Bake 25 minutes, or until the loaves are browned on top and bottom and sound hollow when tapped. Remove the baking sheets and cool on wire racks.

MAKES 4 LOAVES (ABOUT 6 SERVINGS EACH)

Calories:	61	*Total fat:*	0.2 g
Protein:	2.0 g	*Saturated fat:*	0
Carbohydrates:	12.7 g	*Cholesterol:*	0
Fiber:	1.0 g	*Sodium:*	136 mg

WHEAT BERRY BREAD

Wheat berries add a denseness to the texture of this hearty bread. For a nice finishing touch, brush the top of the loaf with beaten egg white before baking — it makes the baked crust shiny.

In a glass measuring cup, stir together the warm water and sugar. Stir in the yeast and let stand until ¼ inch of bubbly foam forms on top.

In a large bowl, combine the whole wheat flour, milk powder, and salt.

In a medium bowl, stir together

½ cup very warm water (105°–115° F)
½ teaspoon sugar
1 package active dry yeast
1½ cups whole wheat flour
¼ cup nonfat dry milk powder
2 teaspoons salt
3 tablespoons molasses
3 tablespoons honey
3 tablespoons vegetable oil
1 cup cooked wheat berries (whole-grain wheat)
1¼ to 1¾ cups all-purpose flour

the yeast mixture, molasses, honey, and oil until combined. Add the wheat berries. Stir this liquid mixture into the flour mixture. Mix in 1¼ cups of the all-purpose flour to make a dough that is easy to handle.

Turn the dough out onto a well-floured surface and knead in enough of the remaining all-purpose flour to make a dough that is smooth, elastic, and no longer sticky. Place the dough in a large greased bowl, turn to coat all over with the grease, and cover with greased plastic wrap. Set in a warm, draft-free spot until doubled in bulk.

Punch the dough down and form into a loaf. Place in a greased 9 × 5 × 2¾-inch loaf pan. Cover with greased plastic wrap and let rise until doubled in bulk.

Preheat the oven to 350° F. Bake for 40 minutes, or until the loaf is browned on top and bottom and sounds hollow when tapped. Cool on a wire rack.

MAKES 1 LOAF (14 SLICES)

Calories:	163	*Total fat:*	3.4 g
Protein:	4.6 g	*Saturated fat:*	0.2 g
Carbohydrates:	26.5 g	*Cholesterol:*	0
Fiber:	2.0 g	*Sodium:*	237 mg

❦ RED PEPPER PIZZA BREAD (V)

This bread is a vivid orange, and its flavor is certainly more interesting than that of any plain pizza dough I've ever tasted. The red pepper and tomato paste take the place of the tomato sauce of a traditional pizza, and we skip the cheese topping.

For the red pepper puree: Cut the bell pepper in quarters and discard the stem, pith, and seeds. Broil on both sides until charred. Place in a paper bag or wrap in foil and let stand until cooled. Or, to prepare in a microwave oven, place the pepper quarters on waxed paper in the microwave and cook on high (100 percent) power 2 minutes. Turn over and cook on high power 1 minute longer. Place in a paper bag and let stand 10 minutes. Peel the pepper, discarding the skin, and place in a blender container or the workbowl of a food processor fitted with a steel blade. Add the 3 tablespoons water and the tomato paste, garlic, and red pepper. Cover and process until smooth.

For the dough: In a glass measuring cup, stir together the warm water and sugar. Stir in the yeast and let stand until ¼ inch of bubbly foam forms on top.

In a large bowl, combine the semolina flour, whole wheat flour, and salt. Stir in the yeast mixture, ¾ cup water, red pepper puree, chopped sun-dried tomatoes, and the 1 ta-

RED PEPPER PUREE
1 red bell pepper
3 tablespoons water
1 tablespoon tomato paste
2 cloves garlic, minced
⅛ teaspoon ground red pepper

DOUGH
½ cup very warm water (105°–115° F)
½ teaspoon sugar
1 package active dry yeast
1 cup semolina flour
1 cup whole wheat flour
1½ teaspoons salt
¾ cup water
½ cup chopped sun-dried tomatoes
1 tablespoon olive oil
2¾ to 3¼ cups all purpose flour

TOPPINGS
2 tablespoons olive oil
2 cups thinly sliced onion
½ teaspoon dried rosemary, crumbled
½ teaspoon dried basil
¼ teaspoon dried thyme

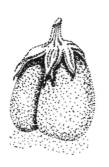

blespoon oil. Stir in 1 cup of the all-purpose flour to make a dough that is easy to handle.

Turn the dough out onto a well-floured surface and knead in enough of the remaining all-purpose flour to make a dough that is smooth, elastic, and no longer sticky. Place the dough in a large greased bowl, turn to coat all over with the grease, and cover with greased plastic wrap. Set in a warm, draft-free spot until doubled in bulk.

For the toppings: While the dough is rising, heat the 2 tablespoons oil in a medium skillet. Add the onion and cook, stirring, until softened. Remove from the heat and stir in the rosemary, basil, and thyme.

To assemble: Preheat the oven to 400° F. Punch the dough down, divide in half, and place each half in a 9-inch-round greased baking pan. Press the dough down until it covers the bottom of the pan and reaches about ½ inch up the sides. Cover and let stand 20 minutes. Punch down the bottom of the dough, leaving the sides puffy. Top each bread with half of the onion-herb mixture. Drizzle any oil left in the skillet over each of the breads.

Bake for 30 minutes, or until the onions are browned and the breads are browned on top and bottom and sound hollow when tapped. Cool on wire racks. Serve cut into thin wedges.

MAKES 2 BREADS (8 SERVINGS EACH)

Calories:	336	*Total fat:*	7.9 g
Protein:	9.1 g	*Saturated fat:*	1.1 g
Carbohydrates:	64.8 g	*Cholesterol:*	0
Fiber:	8.4 g	*Sodium:*	406 mg

✌ GARLIC BITES (V)

It's only fair to warn you that the smell of this bread will make you so hungry that you may just gobble up four or five "bites" before you even know it. If you don't have a garlic press, mince the garlic as fine as possible. You don't want to let these bites get too brown or they will get hard.

In a glass measuring cup, stir together the warm water and sugar. Stir in the yeast and let stand until ¼ inch of bubbly foam forms on top.

While the yeast is proofing, stir together 1½ cups of the all-purpose flour and the cornmeal and salt in a large bowl. Add the yeast mixture, water, 1 tablespoon of the oil, and the fresh garlic; stir until combined. Stir in ⅓ cup more all-purpose flour to form a manageable dough.

Turn the dough out onto a heavily floured surface and knead about 7 minutes, using as much of the remaining all-purpose flour as necessary to create a dough that is smooth and elastic (it may still be slightly tacky).

Place the dough in a greased bowl, turning once so the top is oiled. Cover with greased plastic

½ cup very warm water (105°–115° F)
½ teaspoon sugar
1 package active dry yeast
2 to 2½ cups all-purpose flour, divided
⅓ cup white cornmeal
2 teaspoons salt
½ cup water
3 to 5 tablespoons olive oil, divided
4 large cloves garlic, put through a garlic press
Garlic salt or powder

wrap and let rise in a warm, draft-free spot until doubled in bulk.

Punch the dough down and pinch off small pieces of it, forming each into a small ball about ¾ inch in diameter. Place the "bites" on greased baking sheets about 1 inch apart. Brush the top of each bite with some of the remaining olive oil and sprinkle with the garlic salt. Let stand, covered with greased plastic wrap, 25 minutes.

Preheat the oven to 400° F and bake 15 to 18 minutes, or until the tops of the bites are just beginning to brown. Remove from the oven and brush the tops again with the oil. Cool on a wire rack.

MAKES 60 BITES

PER BITE:

Calories:	25	Total fat:	0.7 g
Protein:	0.6 g	Saturated fat:	0.1 g
Carbohydrates:	4.0 g	Cholesterol:	0
Fiber:	0.2 g	Sodium:	71 mg

PEPPER-PEAR BREAD (V)

½ cup very warm water
(105°–115° F)
1 teaspoon sugar
2 packages active dry yeast
3 cups graham flour
2 teaspoons salt
2 teaspoons medium-grind
pepper

The combination of slightly spicy and slightly sweet makes this a special bread. I use a medium grind of pepper, and instead of pureeing the pears myself, I use jarred baby food.

In a glass measuring cup, stir together the warm water and sugar.

Stir in the yeast and let stand until ¼ inch of bubbly foam forms on top.

While the yeast is proofing, stir together the graham flour, salt, pepper, and nutmeg in a large bowl. Add the yeast mixture, pear, water, honey, and oil, stirring until combined. Stir in 1 cup of the all-purpose flour. The dough should be soft but manageable.

Turn the dough out onto a heavily floured surface and knead about 7 minutes, using as much of the remaining all-purpose flour as necessary to create a dough that is smooth and elastic (it may still be slightly tacky).

Place the dough in a greased bowl and turn to coat all over with the grease. Cover with greased plastic wrap and let rise in a warm, draft-free spot until doubled in bulk.

Punch the dough down and form into a 6-inch round; place on a greased baking sheet. Using a sharp knife, score the top. Brush with the egg white (if using) and, if desired, sprinkle with additional pepper. Cover with greased plastic wrap and let rise until doubled in bulk.

Preheat the oven to 350° F and bake 15 minutes, or until the loaf is browned on top and bottom and sounds hollow when tapped; cool on a wire rack.

MAKES 1 LOAF (20 SERVINGS)

¼ teaspoon ground nutmeg
¾ cup pureed cooked pear
½ cup water (or milk)
⅓ cup honey
2 tablespoons vegetable oil
2 cups all-purpose flour
1 egg white, beaten (optional)

Calories:	140	*Total fat:*	2.0 g
Protein:	4.0 g	*Saturated fat:*	0.4 g
Carbohydrates:	27.4 g	*Cholesterol:*	0.8 mg
Fiber:	2.7 g	*Sodium:*	217 mg

✌ WALNUT-RAISIN ROLLS   (V)

½ cup very warm water
  (105°–115° F)
½ teaspoon sugar
2 packages active dry yeast
1½ cups whole wheat flour
½ cup quick-cooking oats
½ cup ground walnuts
⅓ cup oat bran flakes or
  oat bran
1½ teaspoons salt
1½ cups raisins
1 cup chopped walnuts
½ cup water
¼ cup pure maple syrup
2 tablespoons walnut or
  vegetable oil
1 to 1½ cups all-purpose
  flour

Don't worry if your dough doesn't rise very much — these rolls will be divine anyway. Be sure to use plump raisins. I buy mine at the health food store.

In a glass measuring cup, stir together the warm water and sugar. Stir in the yeast and let stand until ¼ inch of bubbly foam forms on top.

In a large bowl combine the whole wheat flour, oats, walnuts, oat bran flakes, and salt. Stir in the raisins and walnuts.

Stir in the yeast mixture, water, maple syrup, and oil until combined. Add ½ cup of the all-purpose flour to make a dough that is easy to handle.

Turn the dough out onto a well-floured surface and knead in enough of the remaining all-purpose flour to make a dough that is smooth, elastic, and no longer sticky. Place the dough in a large greased bowl, turn to coat all over with the grease, and cover with greased plastic wrap. Set in a warm, draft-free spot until doubled in bulk.

Punch the dough down and divide into 15 equal pieces. Shape into tight balls. Place on a greased 9 × 13 × 2-inch baking pan about 2 inches apart. Cover with greased plastic wrap and let rise until doubled in bulk.

Preheat the oven to 350° F. Bake

for 40 minutes, or until browned on top. Cool on wire racks.

MAKES 15 ROLLS

Calories:	250	*Total fat:*	9.8 g
Protein:	6.5 g	*Saturated fat:*	0.9 g
Carbohydrates:	36.7 g	*Cholesterol:*	0
Fiber:	3.9 g	*Sodium:*	219 mg

❧ BUTTERMILK BISCUITS (E)

These biscuits are wonderful spread with jam. There is no sugar in the recipe, so if you like your biscuits slightly sweet, stir 2 tablespoons sugar into the flour mixture before cutting in the shortening.

Preheat the oven to 400° F.

In a large bowl, using a whisk, stir together the flours and the baking powder, baking soda, and salt. With a pastry cutter or 2 knives in a scissorlike motion, cut the shortening into the flour until the mixture resembles cornmeal. Stir in the buttermilk (not all the flour will be easily incorporated). Turn the dough and any flour remaining in the bowl out onto a lightly floured surface and knead 12 times. Roll into a 10-inch circle, ½ inch thick. Cut into 2-inch-round biscuits. Place on an ungreased baking sheet and bake 12 to 15 minutes, or until browned.

MAKES 15 BISCUITS

1 cup all-purpose flour
1 cup whole wheat flour
1 teaspoon baking powder
1 teaspoon baking soda
1 teaspoon salt
¼ cup vegetable shortening
¾ cup buttermilk

Calories:	99	*Total fat:*	3.8 g
Protein	2.4 g	*Saturated fat:*	1.0 g
Carbohydrates:	14.3 g	*Cholesterol:*	0
Fiber:	1.0 g	*Sodium:*	232 mg

❧ AMARANTH CORNBREAD (E)(Q)

1 cup yellow cornmeal
½ cup all-purpose flour
⅓ cup amaranth flour
⅓ cup sugar
1 tablespoon baking powder
1 teaspoon baking soda
½ teaspoon salt
2 eggs (or 3 egg whites)
1⅓ cups buttermilk
⅓ cup margarine, melted

A cornbread that is moister than most, this is one of the few cornbreads that don't go stale quickly — and it's delicious.

Preheat the oven to 400° F. Grease an 8 × 8 × 2-inch baking pan.

In a large bowl, stir together the cornmeal, both flours, and the sugar, baking powder, baking soda, and salt.

In a medium bowl, beat the eggs. Stir in the buttermilk and margarine. Add the liquid ingredients to the dry ingredients and blend until just combined. Pour into the prepared pan; bake 25 minutes, or until a wooden pick inserted in the center comes out clean. Cut into 9 squares.

SERVES 9

Calories:	217	*Total fat:*	9.1 g
Protein:	5.2 g	*Saturated fat:*	4.8 g
Carbohydrates:	29.8 g	*Cholesterol:*	65.7 mg
Fiber:	1.8 g	*Sodium:*	430 mg

~ TEFF BANANA BREAD

The bananas should be very ripe to have the desired sweetness; look for ones with lots of small dark specks on the skin. If you can't find teff flour, substitute whole wheat flour.

Preheat the oven to 350° F. Grease and flour a 9 × 5 × 2¾-inch loaf pan.

In a medium bowl or on a piece of waxed paper, stir together the flours and the baking powder, baking soda, and salt with a whisk.

In a large bowl, cream the margarine with both sugars until light and fluffy. Beat in the eggs, lemon juice, and lemon rind until thoroughly combined. Beat in the banana alternately with the flour mixture.

Spoon the batter into the prepared pan. Bake 50 minutes, or until a wooden pick inserted in the center comes out clean. Remove from the oven and turn onto a wire rack to cool.

MAKES 1 LOAF (14 SLICES)

1¼ cups all-purpose flour
¾ cup teff flour
1 tablespoon baking powder
1 teaspoon baking soda
1 teaspoon salt
½ cup margarine, softened
½ cup firmly packed light or dark brown sugar
⅓ cup sugar
2 eggs (or 3 egg whites)
1 tablespoon fresh lemon juice
2 teaspoons grated lemon rind
1½ cups mashed ripe banana (about 3 to 4 medium bananas)

Calories:	202	Total fat:	7.7 g
Protein:	3.3 g	Saturated fat:	0.9 g
Carbohydrates:	31.4 g	Cholesterol:	39 mg
Fiber:	1.6 g	Sodium:	351 mg

CRANBERRY-PUMPKIN BREAD

2½ cups sugar, divided
2 cups chopped fresh or
    frozen cranberries
3½ cups Oat Blend flour or
    all-purpose flour
2 teaspoons baking soda
2 teaspoons baking powder
1 teaspoon salt
½ teaspoon pumpkin pie
    spice or ground cinna-
    mon
1 can (16 ounces) solid-
    pack pumpkin *or* 2 cups
    pureed pumpkin
1 cup vegetable oil
¾ cup buttermilk
1 teaspoon vanilla extract
4 eggs (or 6 egg whites)

This bread is so moist and delicious that it gets gobbled up in the blink of an eye. The recipe makes two large loaves, because I like to make it at holiday time, when there are lots of people to serve. If, by some accident, there is any left over, just slice and freeze.

Preheat the oven to 325° F. Grease and flour two 9 × 5 × 2¾-inch loaf pans.

In a medium bowl, stir together ½ cup of the sugar and the cranberries; set aside.

In a medium bowl or on a piece of waxed paper, stir together the flour, the remaining 2 cups sugar, and the baking soda, baking powder, salt, and pumpkin pie spice; set aside.

In a large bowl, beat the pumpkin, oil, buttermilk, and vanilla until combined. Beat in the eggs. Stir in the flour mixture and then the cranberries.

Spoon into the prepared pans and bake 1 hour, or until a wooden pick inserted in the center comes out clean. Turn onto a wire rack to cool.

MAKES 2 LOAVES (16 SLICES EACH)

Calories:	183	*Total fat:*	7.9 g
Protein:	2.9 g	*Saturated fat:*	0.7 g
Carbohydrates:	26.8 g	*Cholesterol:*	34.5 mg
Fiber:	2.2 g	*Sodium:*	154 mg

❧ HONEY-ORANGE BREAD

The sweetness of the honey and the tang of the orange rind give this bread a nice balance. It's perfect spread with yogurt cheese and a little marmalade.

Preheat the oven to 350° F. Grease a 9 × 5 × 2¾-inch loaf pan.

In a large bowl, using a whisk, stir together both flours and the sugar, baking powder, baking soda, and salt. Using a pastry cutter or 2 knives in a scissorlike motion, cut the margarine into the flour until the mixture resembles coarse cornmeal.

In a medium bowl, beat the juice, honey, eggs, and orange rind until thoroughly combined. Stir into the dry ingredients until just moistened. Turn into the prepared pan. Bake 1 hour, or until a wooden pick inserted in the center comes out clean. Turn onto a wire rack to cool.

MAKES 1 LOAF (16 SLICES)

1¼ cups graham or whole wheat flour
1¼ cups all-purpose flour
¾ cup sugar
2 teaspoons baking powder
1 teaspoon baking soda
½ teaspoon salt
½ cup margarine
1 cup orange juice
½ cup honey
2 eggs (or 3 egg whites)
1 tablespoon grated orange rind

Calories:	232	*Total fat:*	7.9 g
Protein:	4.1 g	*Saturated fat:*	3.8 g
Carbohydrates:	38.7 g	*Cholesterol:*	39.3 mg
Fiber:	1.7 g	*Sodium:*	248 mg

❧ MOLASSES–RAISIN BRAN MUFFINS (E) (Q)

1 cup All-Bran cereal
1 cup warm milk
⅓ cup molasses
2 egg whites
½ cup raisins
1 cup all-purpose flour
2 tablespoons sugar
1 tablespoon baking powder
1 teaspoon baking soda
¼ teaspoon salt

These muffins don't rise very high, but they are extremely moist and tasty. If you don't have All-Bran, you can substitute Bran Buds.

Preheat the oven to 350° F. Grease 12 three-inch muffin cups.

Combine the All-Bran and milk in a medium bowl; let stand 5 minutes. Beat in the molasses, then the egg whites and raisins.

In a medium bowl or on a piece of waxed paper, stir together the flour, sugar, baking powder, baking soda, and salt. Add the bran mixture and stir until just combined.

Divide the batter evenly among the prepared cups and bake 12 to 15 minutes, or until a wooden pick inserted in the center comes out clean. Cool on a wire rack.

MAKES 12 MUFFINS

Calories:	118	*Total fat:*	1.0 g
Protein:	3.5 g	*Saturated fat:*	0.5 g
Carbohydrates:	26.3 g	*Cholesterol:*	2.8 mg
Fiber:	2.6 g	*Sodium:*	190 mg

ZUCCHINI-BANANA BRAN MUFFINS

The flavor of banana is distinct in these muffins but not overwhelming. If you like, you can stir in some raisins. Since the muffins do not rise too high, be sure to fill the muffin cups at least three quarters full.

Preheat the oven to 350° F. Thoroughly grease 12 two-and-a-half-inch muffin cups.

Combine the All-Bran and milk in a medium bowl; let stand 5 minutes. Beat in the banana and zucchini, then the egg whites, sugar, and oil.

In a medium bowl or on a piece of waxed paper, stir together both flours and the baking powder and salt. Add the bran mixture and stir until just combined.

Divide the batter evenly among the prepared muffin cups, filling them three quarters full. Bake 25 to 30 minutes, or until lightly browned on top. Cool on a wire rack.

MAKES 12 MUFFINS

¾ cup All-Bran cereal
½ cup warm milk
¾ cup very ripe mashed banana (2 small bananas)
1 cup shredded zucchini
3 egg whites
3 tablespoons sugar
3 tablespoons vegetable oil
1 cup all-purpose flour
⅓ cup graham or whole wheat flour
1 tablespoon baking powder
¼ teaspoon salt

Calories:	130	Total fat:	4.0 g
Protein:	3.7 g	Saturated fat:	0.7 g
Carbohydrates:	21.8 g	Cholesterol:	1.4 mg
Fiber:	2.7 g	Sodium:	177 mg

✍ CORN MUFFINS (E)

1½ cups white or yellow
    cornmeal
1 cup all-purpose flour
⅓ cup sugar
4 teaspoons baking powder
1 teaspoon salt
2 egg whites
1 cup milk
⅓ cup vegetable oil

These corn muffins are dense and slightly sweet. You can stir in some blueberries for a more festive muffin. Although either white or yellow cornmeal will work well, yellow is more traditional for this recipe.

Preheat the oven to 425° F. Thoroughly grease 12 two-and-a-half-inch muffin cups.

In a large bowl, stir together the cornmeal, flour, sugar, baking powder, and salt.

In a medium bowl, lightly beat the egg whites. Beat in the milk and oil until completely combined. Stir the liquid ingredients into the dry ingredients until just moistened.

Divide the batter evenly among the prepared muffin cups (they will be fairly full). Bake 20 to 25 minutes, or until lightly browned on top.

MAKES 12 MUFFINS

Calories:	192	*Total fat:*	7.0 g
Protein:	3.7 g	*Saturated fat:*	1.3 g
Carbohydrates:	28.3 g	*Cholesterol:*	2.7 mg
Fiber:	0.8 g	*Sodium:*	306 mg

↶ WHOLE WHEAT MUFFINS (E)

A rough, craggy-looking top, moist texture, and delicious flavor make these delightful muffins a breakfast favorite.

Preheat the oven to 400° F. Thoroughly grease 12 two-and-a-half-inch muffin cups.

In a large bowl, stir together both flours and the brown sugar, wheat germ, baking powder, baking soda, and salt.

In a medium bowl, beat the egg whites lightly. Add the buttermilk and shortening; beat until combined. Add the liquid ingredients to the flour mixture and stir (do not beat) until just moistened.

Divide the batter evenly among the prepared muffin cups (they will be fairly full). Bake 20 to 25 minutes, or until lightly browned on top.

MAKES 12 MUFFINS

1 cup whole wheat flour
1 cup all-purpose flour
¼ cup firmly packed light or dark brown sugar
3 tablespoons honey-toasted wheat germ
1 tablespoon baking powder
1 teaspoon baking soda
1 teaspoon salt
2 egg whites
1½ cups buttermilk
¼ cup melted vegetable shortening

Calories:	147	*Total fat:*	5.0 g
Protein:	4.3 g	*Saturated fat:*	1.3 g
Carbohydrates:	21.8 g	*Cholesterol:*	1.1 mg
Fiber:	1.7 g	*Sodium:*	371 mg

❧ OLIVE CREAMY CHEESE   (E)

1 cup cottage cheese
½ cup pimento-stuffed
   olives

My favorite lunch, as a kid, was cream cheese and olive sandwiches. Nowadays, cream cheese is too high in fat to consider, but this spread is a perfectly fine substitute.

Place the cottage cheese in the workbowl of a food processor fitted with a steel blade. Cover and process until smooth. Spoon the cottage cheese into the drip basket of a Melitta R coffeemaker fitted with a filter, or into a strainer fitted with a triple layer of cheesecloth. Place the drip basket or strainer over a cup or bowl to catch the drippings and refrigerate 4 hours.

Put the drained cheese and olives into a food processor workbowl fitted with a steel blade. Cover and process until the olives are finely chopped.

MAKES 1 CUP

PER TABLESPOON:

Calories:	22	*Total fat:*	1.7 g
Protein:	1.7 g	*Saturated fat:*	0.6 g
Carbohydrates:	0.5 g	*Cholesterol:*	1.9 mg
Fiber:	0.2 g	*Sodium:*	93 mg

❧ HERB AND GARLIC CREAMY CHEESE (E)

Feel free to vary the herbs you use in this spread. Fresh basil and parsley are nice in addition to the scallion.

Place the cottage cheese in the workbowl of a food processor fitted with a steel blade. Cover and process until smooth. Spoon the cottage cheese into the drip basket of a Melitta R coffeemaker fitted with a filter, or into a strainer fitted with a triple layer of cheesecloth. Place the drip basket or strainer over a cup or bowl to catch the drippings and refrigerate 4 hours.

Put the drained cheese, scallion, garlic, thyme, and marjoram into a food processor workbowl fitted with a steel blade. Cover and process until the scallion is finely chopped.

MAKES 1 CUP

1 cup cottage cheese
1 scallion, cut into 1-inch pieces
1 clove garlic, minced
¼ teaspoon dried thyme
⅛ teaspoon dried marjoram or oregano

PER TABLESPOON:

Calories:	14	*Total fat:*	0.6 g
Protein:	2.3 g	*Saturated fat:*	0.1 g
Carbohydrates:	0.5 g	*Cholesterol:*	0
Fiber:	0	*Sodium:*	53 mg

⌇ LIPTAUER   (E)(Q)

1 container (8 ounces) cottage cheese
3 tablespoons grated Par-
mesan (optional)
1 tablespoon Dijon mus-
tard
1 tablespoon grated onion
or onion juice
2 teaspoons caraway seeds
1 teaspoon capers
1 teaspoon paprika
¼ teaspoon Worcestershire
sauce
⅛ teaspoon ground red
pepper (optional)

This is a modification of an old Hungarian recipe that has been in my family for generations. The original recipe uses anchovies and butter in addition to the ingredients listed here. Be sure to grate the onion on the fine side of the gra-ter — you will get an onion puree and lots of onion juice.

Place all the ingredients in the workbowl of a food processor fitted with a steel blade. Cover and pro-cess until completely combined and fairly smooth.

MAKES 1 CUP

PER TABLESPOON:

Calories:	21	*Total fat:*	1.0 g
Protein:	2.3 g	*Saturated fat:*	0.6 g
Carbohydrates:	0.8 g	*Cholesterol:*	2.8 mg
Fiber:	0	*Sodium:*	87 mg

⌇ SESAME SPREAD   (E)(Q)(V)

1⅓ cups (8 ounces) tofu
⅓ cup tahini (sesame paste)
¼ teaspoon salt
¼ teaspoon sugar

A mildly flavored spread with the consistency of whipped cream cheese, this recipe is very adapta-ble. You can vary the flavor to suit your own tastes by using cashew or peanut butter instead of the tahini. You can also stir in additional fla-vorings — such as brown sugar and cinnamon, soy sauce and chopped scallion, or minced garlic and cumin — depending on the type of bread you intend to serve it with.

Break the tofu into pieces and place in the workbowl of a food processor fitted with a steel blade (this recipe is not well suited to a blender). Add the tahini, salt, and sugar; cover and process until smooth.

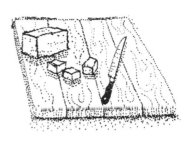

MAKES 1¼ CUPS

PER 2 TABLESPOONS:

Calories:	65	*Total fat:*	5.5 g
Protein:	3.2 g	*Saturated fat:*	0.7 g
Carbohydrates:	1.2 g	*Cholesterol:*	0
Fiber:	1.2 g	*Sodium:*	57 mg

❧ TOFU FRUIT CHEESE (E)(V)

This spread is an ideal protein booster if you are eating muffins or bread for breakfast. You can also eat it before you drain it, when the consistency will be similar to that of a fruit yogurt's.

Place the preserves, honey, and ginger in a blender container or the workbowl of a food processor fitted with a steel blade. Cover and process until smooth. With the machine running, drop in the cubes of tofu until all the tofu has been incorporated.

Place the mixture in the drip basket of a Melitta R type coffeemaker fitted with a filter or in a strainer

¼ cup apricot (or any other flavor) preserves
1 tablespoon honey
¼ teaspoon ground ginger
1⅓ cups (8 ounces) tofu, cut into 1-inch cubes

lined with a quadruple layer of cheesecloth. Place the drip basket or strainer over a cup or bowl and put into the refrigerator overnight for a soft cheese; leave in 24 hours for a firmer cheese.

MAKES ½ CUP

PER TABLESPOON:

Calories:	55	*Total fat:*	1.2 g
Protein:	2.3 g	*Saturated fat:*	0.2 g
Carbohydrates:	9.9 g	*Cholesterol:*	0
Fiber:	0.6 g	*Sodium:*	3 mg

❧ HONEY-ORANGE PEANUT BUTTER (E)(Q)(V)

1 cup roasted peanuts (salted, lightly salted, or unsalted, to taste)
3 tablespoons orange juice
2 tablespoons honey
2 teaspoons grated orange rind

You can create whatever texture you like for this spread by adjusting the amount of time you process it. The consistency is less creamy than that of standard prepared brands of peanut butter.

Place the peanuts, orange juice, honey, and orange rind in the workbowl of a food processor fitted with a steel blade. Cover and process to the desired consistency.

MAKES ¾ CUP

PER TABLESPOON:

Calories:	82	*Total fat:*	5.9 g
Protein:	3.3 g	*Saturated fat:*	0.8 g
Carbohydrates:	5.6 g	*Cholesterol:*	0
Fiber:	1.0 g	*Sodium:*	118 mg

～ HUMMOS (E)(Q)(V)

Hummos is a popular dish throughout the Middle East. You can serve it as a sandwich in pita bread, topped with salad and tahini dressing, or as a spread with pita wedges.

Place the chickpeas, tahini, lemon juice, olive oil, water, garlic, cumin, salt, and Red Hot sauce in the workbowl of a food processor fitted with a steel blade. Cover and process until smooth.

MAKES 1⅔ CUPS

1½ cups cooked or canned (drained and rinsed) chickpeas
¼ cup tahini (sesame paste)
3 tablespoons fresh lemon juice
2 tablespoons olive oil
2 tablespoons water
2 cloves garlic, minced
½ teaspoon ground cumin
½ teaspoon salt
¼ teaspoon Red Hot sauce

PER ⅓ CUP:

Calories:	245	Total fat:	16.2 g
Protein:	7.9 g	Saturated fat:	2.2 g
Carbohydrates:	18.3 g	Cholesterol:	0
Fiber:	4.1 g	Sodium:	374 mg

～ SPANISH-STYLE BEAN SPREAD

Sofrito is a condiment made of tomatoes, peppers, onions, and garlic. You can use it to enhance the flavor of poultry dishes, eggs, or any recipe that needs a little pep. You can find prepared sofrito in Hispanic grocery stores or in the Spanish section of your supermarket.

Place the beans, yogurt, sofrito, cumin, coriander, salt, and red pepper in the workbowl of a food pro-

1 can (15 ounces) Spanish-style pink beans, rinsed and drained, *or* 1½ cups cooked pink beans
½ cup unflavored yogurt
¼ cup sofrito
¼ teaspoon ground cumin
¼ teaspoon ground coriander
¼ teaspoon salt
¼ teaspoon ground red pepper

cessor fitted with a steel blade. Cover and process until smooth.

MAKES 1¾ CUPS

PER 2 TABLESPOONS:

Calories:	50	*Total fat:*	0.3 g
Protein:	3.8 g	*Saturated fat:*	0
Carbohydrates:	8.9 g	*Cholesterol:*	0
Fiber:	3.2 g	*Sodium:*	164 mg

❧ MOCK CHOPPED LIVER (E)(Q)(V)

1½ tablespoons vegetable oil
½ cup chopped onion
4 cups coarsely chopped mushrooms
½ cup chopped carrot
1 cup crushed Tam Tam or Ritz crackers
½ teaspoon salt
¼ teaspoon pepper
⅛ teaspoon ground ginger

This spread not only tastes like chopped liver, it also looks like it and has the same consistency. Of course, the catch is that you have to like chopped liver, which I do.

In a large skillet, heat the oil over high heat. Add the onion and cook, stirring, until browned on the edges. Add the mushrooms and carrot, and cook, stirring, until the liquid that will be given off by the mushrooms evaporates. Place the vegetable mixture in the workbowl of a food processor fitted with a steel blade. Add the crackers, salt, pepper, and ginger. Cover and process until smooth.

MAKES 1½ CUPS

PER ¼ CUP:

Calories:	77	*Total fat:*	5.5 g
Protein:	1.4 g	*Saturated fat:*	0.9 g
Carbohydrates:	7.8 g	*Cholesterol:*	0
Fiber:	1.5 g	*Sodium:*	243 mg

❧ PROVENÇALE SPREAD (E)(Q)(V)

Serve this as an hors d'oeuvre on toasted French bread rounds or as a sandwich spread on a baguette. If you use imported olives from Greece or Italy, the flavor will be even better. (Unfortunately, you'll probably have to pit the olives yourself, but the taste difference is well worth the effort.) The amount of salt needed will vary according to the type of olive you choose; taste before adding any, then adjust accordingly.

1 large tomato
¼ cup fresh basil
¼ cup chopped parsley
2 cloves garlic
½ cup pitted black olives
1 teaspoon olive oil
¼ teaspoon pepper
   Salt to taste

Peel the tomato by placing it in boiling water for 1 minute and then plunging into ice water. The skin should be loosened enough to peel easily. Cut in half widthwise and squeeze out the seeds. Dice the tomato fine (you should have ¾ cup diced tomato) and set aside.

Place the basil, parsley, and garlic into the workbowl of a food processor fitted with a steel blade. Cover and process until finely minced. Add the olives; cover and process until finely chopped. Spoon into a medium bowl and stir in the oil and pepper. Discarding any liquid that may have accumulated from the diced tomatoes, stir the tomatoes into the olive mixture. Taste and add salt as necessary.

MAKES 1 CUP

PER TABLESPOON:

Calories:	23	Total fat:	2.7 g
Protein:	0.2 g	Saturated fat:	0.4 g
Carbohydrates:	0.8 g	Cholesterol:	0
Fiber:	0.6 g	Sodium:	165 mg

Breakfast Dishes

✎ HONEY–WHOLE WHEAT PANCAKES (E)(Q)

The honey rounds out the flavor of these whole wheat pancakes without overwhelming the slight nuttiness of the wheat. For a no-cholesterol version, use 3 egg whites or an egg substitute instead of the 2 eggs, and skim milk.

1 cup whole wheat flour
¾ cup all-purpose flour
2 teaspoons baking powder
½ teaspoon salt
1¾ cups milk
2 eggs
3 tablespoons honey
½ teaspoon vanilla extract
3 tablespoons melted margarine or vegetable oil

In a large bowl, stir together both flours and the baking powder and salt.

In a separate bowl, beat together the milk, eggs, honey, and vanilla. Beat in the margarine. Stir the milk mixture into the dry ingredients until just blended. The mixture should still be lumpy.

Heat a large skillet over medium-high heat until a few drops of water dance across the surface before evaporating. Lightly grease the skillet or spray with no-stick cooking spray. Allowing 2 very full tablespoons of batter per pancake, drop the batter onto the skillet.

Cook until bubbles form on the surface of the pancake. If the bottom of the pancake gets too dark before bubbles form, lower the heat a little. Turn with a wide spatula and cook until the other side is browned.

MAKES 18 FOUR-INCH PANCAKES

Calories:	95	Total fat:	3.2 g
Protein:	2.9 g	Saturated fat:	1.3 g
Carbohydrates:	12.9 g	Cholesterol:	1.2 mg
Fiber:	0.8 g	Sodium:	137 mg

✌ OAT BRAN PANCAKES WITH STRAWBERRY-ORANGE SAUCE (E)(Q)

SAUCE
  1 cup chopped fresh straw-
    berries
  ¾ cup orange juice
  3 tablespoons sugar
  2 tablespoons water
  1 tablespoon cornstarch

PANCAKES
  1 cup all-purpose flour
  ½ cup oat bran
  2 tablespoons sugar
  2 teaspoons baking powder
  1 teaspoon baking soda
  ½ teaspoon salt
  3 egg whites
  2 tablespoons vegetable oil
  1½ cups buttermilk

These pancakes are thick and satisfying. Be sure that the heat is not too high or they will be overdone on the outside while undercooked inside.

For the sauce: In a medium saucepan, combine the strawberries, orange juice, and sugar. Bring to a boil and simmer 3 minutes. In a small bowl, stir together the water and cornstarch until smooth. Stir into the strawberry mixture and bring to a boil. Simmer 1 minute; set aside.

For the pancakes: In a large bowl, stir together the flour, oat bran, sugar, baking powder, baking soda, and salt.

In a medium bowl, mix the egg whites and oil until combined. Stir in the buttermilk. Stir the liquid ingredients into the dry ingredients until just blended.

Heat a skillet over medium heat until a drop of water dances across the surface before evaporating. Grease the skillet lightly or spray with no-stick cooking spray. Allowing 2 very full tablespoons of batter per pancake, drop the batter onto the skillet. Cook until bubbles burst on the top and the pancake is browned on the bottom. Turn and cook until browned on the other side.

MAKES 16 THREE-INCH PANCAKES AND 1¼ CUPS SAUCE

PER PANCAKE:

Calories:	66	*Total fat:*	2.0 g
Protein:	2.3 g	*Saturated fat:*	0.4 g
Carbohydrates:	9.6 g	*Cholesterol:*	0.8 mg
Fiber:	0.5 g	*Sodium:*	197 mg

PER TABLESPOON SAUCE:

Calories:	15	*Total fat:*	0
Protein:	0.1 g	*Saturated fat:*	0
Carbohydrates:	3.8 g	*Cholesterol:*	0
Fiber:	0.2 g	*Sodium:*	1 mg

❧ GERMAN APPLE PANCAKES (Q)

Although these are called pancakes and are shaped like pancakes, they have a dense, chewy texture, unlike regular pancakes, which should be light and fluffy. Traditionally they are served with cinnamon and sugar, but you can also top them with maple or fruit

2 tablespoons margarine, divided
3 cups peeled, thinly sliced apples
3 egg whites
½ cup whole wheat flour
1 cup milk, divided
¾ cup all-purpose flour

3 tablespoons sugar
½ teaspoon baking powder
½ teaspoon salt

syrup. Either way they are delicious. If you don't use a nonstick skillet, you may need a little extra margarine.

Melt 1 tablespoon of the margarine in a large skillet over medium-high heat. Add the apples and cook, stirring, until softened. Cool.

In a large bowl, beat the egg whites until frothy. Stir in the whole wheat flour, then ½ cup of the milk. Add the all-purpose flour, sugar, baking powder, and salt. Mix in the remaining ½ cup milk. Let the batter stand 5 minutes. Stir in the apples.

Melt half the remaining 1 tablespoon margarine in a 10-inch slope-sided nonstick skillet. Drop ¼ cup batter per pancake into the skillet. Cook until browned on the bottom. Turn with a spatula and cook until browned on the other side. Use the remaining margarine, and add more if necessary, to cook the remaining batter.

MAKES 12 PANCAKES

Calories:	78	*Total fat:*	2.8 g
Protein:	2.3 g	*Saturated fat:*	0.8 g
Carbohydrates:	11.9 g	*Cholesterol:*	2.8 mg
Fiber:	1.4 g	*Sodium:*	134 mg

❧ CINNAMON–COTTAGE CHEESE PANCAKES (E)(Q)

Be sure the heat is not too high and use a nonstick skillet, since these pancakes are quite delicate and burn easily.

Place the cottage cheese, egg whites, sugar, vanilla, cinnamon, and salt in a blender container. Cover and blend at high speed until smooth. Pour the mixture into a bowl. Add both flours and the baking powder and stir until combined.

In a large skillet, over medium heat, melt 1 teaspoon of the margarine. Drop 4 pancakes (3 tablespoons of batter per pancake) onto the skillet.

Cook until bubbles burst on the surface. Turn and cook until lightly browned on the other side. Repeat with the remaining margarine and batter.

MAKES 8 PANCAKES

⅔ cup cottage cheese
4 egg whites
3 tablespoons sugar
1 teaspoon vanilla extract
1 teaspoon ground cinnamon
¼ teaspoon salt
½ cup whole wheat flour
¼ cup all-purpose flour
1½ teaspoons baking powder
2 teaspoons margarine, divided

Calories:	92	Total fat:	2.3 g
Protein:	6.1 g	Saturated fat:	0.8 g
Carbohydrates:	10.1 g	Cholesterol:	2.6 mg
Fiber:	1.3 g	Sodium:	241 mg

❧ ORANGE-GRAHAM WAFFLES (Q)

1 cup graham flour
1 cup all-purpose flour
1 tablespoon baking pow-
der
½ teaspoon salt
1 cup milk
⅓ cup thawed orange juice
concentrate
⅓ cup vegetable oil
3 tablespoons honey
3 egg whites

The graham flour gives these waffles a slightly crunchy texture. If you cannot find graham flour, use 1 cup minus 1 tablespoon whole wheat flour and stir in 1 tablespoon wheat germ.

Preheat a waffle iron as manufacturer directs.

In a large bowl, stir together both flours and the baking powder and salt. In a medium bowl, stir together the milk, orange juice concentrate, oil, and honey. Add the liquid ingredients to the dry ingredients and stir until just blended. (The batter should be lumpy.)

In a clean, grease-free medium bowl, beat the egg whites with clean beaters until stiff peaks form when the beaters are lifted. Fold the egg whites into the batter.

Spread half the batter onto the prepared waffle iron. Bake until the steaming stops, about 5 minutes, or until the waffle is browned. Repeat with the remaining batter.

MAKES 8 FOUR-INCH-SQUARE WAFFLES

Calories:	257	Total fat:	10.0 g
Protein:	6.2 g	Saturated fat:	1.2 g
Carbohydrates:	35.5 g	Cholesterol:	4.1 mg
Fiber:	2.4 g	Sodium:	292 mg

∾ PUMPKIN PIE WAFFLES (E)(Q)

Because I prepare these waffles with canned pumpkin, I can make them year-round and have a little bit of Thanksgiving anytime at all. If you don't have pumpkin pie spice on hand, substitute ¼ teaspoon ground cinnamon, ¼ teaspoon ground ginger, ⅛ teaspoon ground nutmeg, and a pinch of ground cloves and allspice for the 2 teaspoons called for.

Preheat a waffle iron as manufacturer directs.

In a large bowl, stir together both flours and the oat bran, brown sugar, baking powder, pumpkin pie spice, and salt.

In a medium bowl, stir together the milk, pumpkin, oil, and honey until smooth. Add to the dry ingredients.

In a clean, grease-free medium bowl, with clean beaters, beat the egg whites until stiff but not dry. Fold into the pumpkin mixture.

Pour about 1¼ cups of the batter onto the prepared waffle iron. Bake until golden and crisp.

MAKES 12 FOUR-INCH-SQUARE WAFFLES

1 cup all-purpose flour
½ cup whole wheat flour
⅓ cup oat bran
3 tablespoons firmly packed light or dark brown sugar
1 tablespoon baking powder
2 teaspoons pumpkin pie spice
¼ teaspoon salt
1⅓ cups milk
¾ cup pureed pumpkin or canned solid-pack pumpkin
⅓ cup vegetable oil
3 tablespoons honey
3 egg whites

Calories:	166	*Total fat:*	7.2 g
Protein:	3.8 g	*Saturated fat:*	1.4 g
Carbohydrates:	22.6 g	*Cholesterol:*	3.7 mg
Fiber:	1.5 g	*Sodium:*	120 mg

ಱ CORNMEAL WAFFLES (Q)

1¼ cups all-purpose flour
¾ cup white or yellow
  cornmeal
¼ cup sugar
1 tablespoon baking pow-
  der
1 teaspoon baking soda
½ teaspoon salt
4 egg whites
1⅔ cups buttermilk
¼ cup vegetable oil

These waffles have the slightly gritty texture you look for in any cornmeal baked good, and they make a very satisfying breakfast.

Preheat a waffle iron as manufacturer directs.

In a large bowl, stir together the flour, cornmeal, sugar, baking powder, baking soda, and salt.

In a medium bowl, beat the egg whites lightly. Stir in the buttermilk and oil. Add to the dry ingredients and stir until just blended. (The batter should be lumpy.)

Spread 1¼ cups of the batter onto the prepared waffle iron. Bake until the steaming stops, about 5 minutes, or until the waffle is browned. Repeat with the remaining batter.

MAKES 12 FOUR-INCH-SQUARE
WAFFLES

Calories:	155	*Total fat:*	5.0 g
Protein:	4.4 g	*Saturated fat:*	0.9 g
Carbohydrates:	22.7 g	*Cholesterol:*	13 mg
Fiber:	0.6 g	*Sodium:*	292 mg

❧ WHOLE WHEAT CREPES (E)(Q)

A great brunch dish for company, these crepes are delicious topped with sweetened unflavored yogurt and fresh berries. The new, excellent no-sugar-added fruit syrups also make perfect toppings.

Place the milk, both flours, and the egg whites, margarine, sugar, vanilla, and salt in a blender container or the workbowl of a food processor fitted with a steel blade. Cover and process until smooth. Let stand 30 minutes.

Place a slope-sided 6-inch skillet over medium heat. When the skillet is hot enough to make a drop of water sizzle as it evaporates, brush lightly with the oil. Pour about 3 tablespoons of the batter into the skillet and immediately tilt and rotate the pan, so that the batter covers the bottom completely. Cook until the top of the crepe is no longer shiny and the bottom is browned. Using a spatula, carefully turn the crepe and cook the other side about 30 seconds, or until brown spots appear. Turn onto a serving plate and fold into quarters. (If you are preparing these in advance, turn onto waxed paper and stack with a sheet of waxed paper between each crepe. Crepes can be frozen for future use or refrigerated if you are planning on using them within 2 or 3 days.)

Serve with your topping of choice.

MAKES 20 CREPES

1½ cups milk
1 cup whole wheat flour
¼ cup all-purpose flour
3 egg whites
3 tablespoons melted margarine or vegetable oil
2 tablespoons sugar
1 teaspoon vanilla extract
¼ teaspoon salt
Oil for greasing the skillet

Calories:	66	*Total fat:*	3.0 g
Protein:	2.2 g	*Saturated fat:*	0.9 g
Carbohydrates:	8.0 g	*Cholesterol:*	29.9 mg
Fiber:	0.8 g	*Sodium:*	36 mg

❧ CHEESE BLINTZES

1 package (7½ ounces)
   farmer cheese
½ cup cottage cheese
¼ cup confectioners' sugar
8 Whole Wheat Crepes
   (see preceding recipe)
2 teaspoons margarine

Blintzes are a perfect use for any leftover crepes. You can make them in advance and freeze them for future use. Serve them with applesauce, unflavored yogurt, or cinnamon and sugar. If you can't find farmer cheese, substitute an equal amount of ricotta with 1 tablespoon flour stirred in.

In a medium bowl, stir together the farmer cheese, cottage cheese, and confectioners' sugar; set aside.

Place 1 of the crepes on a work surface. Shape 2 very full tablespoons of the cheese mixture into a log and set in the center of the crepe (the log should be parallel to the work surface facing you). Fold the left and right sides of the crepe in toward the center, over the filling. Take the side of the crepe nearest you and fold it over the filling. Roll the crepe to make a package that completely encloses the cheese filling. Repeat with all the other crepes and filling.

Melt the margarine in a large skillet, over medium heat. Add the blintzes, folded flap down, and cook until the bottoms are browned. Turn and cook until browned.

MAKES 8 BLINTZES

Calories:	255	Total fat:	14.9 g
Protein:	12.1 g	Saturated fat:	6.5 g
Carbohydrates:	17.1 g	Cholesterol:	82 mg
Fiber:	1.8 g	Sodium:	166 mg

✒ ORANGE SPOONBREAD

I like to serve this dish for Sunday brunch. When I'm having company, I top it with this easy raspberry sauce: In a blender, process a 10-ounce package of frozen raspberries in light syrup, thawed, with 2 teaspoons of cornstarch and, if desired, 1 tablespoon of raspberry or orange liqueur. Place in a small saucepan and bring to a boil, stirring. Serve the sauce warm. If you prefer to serve it cool, reduce the cornstarch to 1½ teaspoons.

1 cup yellow cornmeal
¼ cup sugar
½ teaspoon salt
1½ cups scalded milk
2 tablespoons margarine
1 egg
1 cup orange juice
2 teaspoons grated orange rind
2 teaspoons baking powder
½ teaspoon baking soda
⅛ teaspoon ground nutmeg
3 egg whites

Preheat the oven to 350° F. Grease a 2-quart casserole.

In a 2-quart saucepan, stir together the cornmeal, sugar, and salt. Gradually add the scalded milk. Cook over low heat, stirring constantly with a whisk, until the mixture is very thick; remove from the heat. Stir in the margarine until completely melted.

In a large bowl, lightly beat the whole egg. Stir in the orange juice, orange rind, baking powder, baking soda, and nutmeg. Whisk in one third of the cornmeal mixture at a time until completely combined and lump-free.

In a clean, grease-free medium bowl, beat the egg whites with clean beaters until stiff but not dry. Fold the whites into the cornmeal mixture. Pour into the prepared baking dish and bake 35 minutes, or until well browned on top.

SERVES 4 TO 6

Calories:	341	Total fat:	10.6 g
Protein:	10.3 g	Saturated fat:	3.5 g
Carbohydrates:	51.2 g	Cholesterol:	80.9 mg
Fiber:	1.3 g	Sodium:	506 mg

❧ GRANOLA (E)(Q)(V)

¼ cup margarine
¼ cup firmly packed light or dark brown sugar
2 tablespoons pure maple syrup
1 tablespoon molasses
2 cups old-fashioned rolled oats
⅔ cup sliced almonds
⅓ cup oat bran
⅓ cup sunflower seeds
⅓ cup pumpkin seeds
⅓ cup wheat germ
¾ cup raisins

Homemade granola is much better than store-bought. You can always tailor it to your own tastes, adding fruits or nuts that you like and leaving out those you're not too fond of.

Preheat the oven to 350° F.

Place the margarine in a 9 × 5 × 2-inch baking dish. Let stand in the oven 5 minutes to melt. Remove from the oven and stir in the sugar, maple syrup, and molasses until combined. Stir in the oats, almonds, oat bran, sunflower seeds, and pumpkin seeds. Bake 15 minutes, stirring twice. Mix in the wheat germ; bake 10 minutes longer, stirring once. Stir in the raisins.

SERVES 12

Calories:	217	Total fat:	10.2 g
Protein:	5.9 g	Saturated fat:	2.2 g
Carbohydrates:	28.6 g	Cholesterol:	0
Fiber:	4.2 g	Sodium:	38 mg

~ MUESLI (E)(Q)(V)

Muesli is a breakfast cereal orig-inally formulated in Switzerland. You can eat it with milk, like reg-ular breakfast cereal, or soak it overnight in water and eat it cold or warm the next morning.

Toss all the ingredients together. Store in an airtight container.

MAKES EIGHT ½-CUP SERVINGS

2 cups old-fashioned rolled oats
½ cup oat bran
½ cup raisins
½ cup chopped dried ap-ples
⅓ cup chopped dried apri-cots
¼ cup sunflower seeds
1 teaspoon ground cinna-mon

Calories:	161	*Total fat:*	3.7 g
Protein:	5.2 g	*Saturated fat:*	0.5 g
Carbohydrates:	29.8 g	*Cholesterol:*	0
Fiber:	5.0 g	*Sodium:*	6 mg

~ EXTRA-HIGH-FIBER OATMEAL (E)(Q)(V)

Oat bran is a great source of fiber, but it's not too tasty when eaten on its own. Here's my solution: oat-meal with extra bran stirred in. If you're not a vegan, use the optional milk — it makes the cereal taste richer.

In a 1-quart saucepan, bring the water, milk, and salt to a boil. Stir in the oats and oat bran. Cook over medium heat, stirring occasionally, 5 minutes, or until the oatmeal reaches the desired consistency.

SERVES 2

1½ cups water
⅓ cup soy milk (or dairy milk)
⅛ teaspoon salt
⅔ cup old-fashioned rolled oats
⅓ cup oat bran

Calories:	175	Total fat:	6.2 g
Protein:	7.9 g	Saturated fat:	2.0 g
Carbohydrates:	28.6 g	Cholesterol:	5.5 mg
Fiber:	3.7 g	Sodium:	154 mg

✤ CRAZY MIXED-UP CEREAL (E)(Q)(V)

1 cup quick-cooking
  Cream of Wheat or fa-
  rina
1 cup Maltex
1 cup oat bran
¾ cup water
¼ cup soy milk (or dairy
  milk)
⅛ teaspoon salt

I tend to keep a number of different cereals on hand at all times. Sometimes I make my own morning mixtures. This one is another way of sneaking some extra oat bran into my breakfast.

In a medium bowl, stir together the Cream of Wheat, Maltex, and oat bran.

In a 1-quart saucepan, bring the water, milk, and salt to a boil (the amounts given opposite are per serving). Gradually stir in ¼ cup of the cereal mix (place the remainder in a storage container for later use; this amount will yield 11 more servings). Cook over medium heat, stirring frequently, about 3 minutes, or until thickened.

SERVES 1

Calories:	124	Total fat:	2.2 g
Protein:	4.2 g	Saturated fat:	0
Carbohydrates:	26.1 g	Cholesterol:	0
Fiber:	1.4 g	Sodium:	3 mg

Bibliography

Atlas, Nava. *The Wholefood Catalog: A Complete Guide to Natural Foods*. New York: Fawcett Columbine, 1988.

Baird, Pat. *Quick Harvest: A Vegetarian's Guide to Microwave Cooking*. New York: Prentice Hall Press, 1991.

Brody, Jane. *Jane Brody's Good Food Book*. New York: W. W. Norton, 1985.

Dried Beans and Grains. Alexandria, Va.: Time-Life Books, 1982.

Farwagi, Peta Lyn. *Full of Beans: An International Bean Cookbook*. New York: Harper & Row, 1978.

FitzGibbon, Theodora. *The Food of the Western World: An Encyclopedic Dictionary*. New York: Quadrangle/The New York Times, 1976.

Gelles, Carol. *The Complete Whole Grain Cookbook*. New York: Donald I. Fine, 1989.

Horsley, Janet. *Bean Cuisine*. Dorset, England: Prism Press, 1982.

Lappé, Frances Moore. *Diet for a Small Planet*. New York: Ballantine Books, 1971.

McGee, Harold J. *On Food and Cooking: The Science and Lore of the Kitchen*. New York: Scribner's, 1984.

National Research Council. *Recommended Dietary Allowances*, 10th ed. Washington, D.C.: National Academy Press, 1989.

Netzer, Corinne T. *The Complete Book of Food Counts*. New York: Dell, 1988.

Physicians Committee for Responsible Medicine. "Guide to Healthy Eating." Washington, D.C., November–December 1990.

Roehl, Evelyn. *Whole Food Facts*. Rochester, Vt.: Healing Arts Press, 1988.

Root, Waverly. *Food.* New York: Simon & Schuster, 1980.

Sass, Lorna J. *Cooking Under Pressure.* New York: William Morrow, 1989.

Scott, Maria Luisa, and Jack Denton Scott. *The Bean, Pea, and Lentil Cookbook.* New York: Consumer Reports Books, 1991.

Stone, Sally, and Martin Stone. *The Brilliant Bean: Sophisticated Recipes for the World's Healthiest Food.* Toronto: Bantam Books, 1988.

United States Department of Agriculture. *Agricultural Handbook no. 8, Composition of Foods.* Washington, D.C.: U.S. Government Printing Office, 1975.

White, Beverly. *Bean Cuisine: A Culinary Guide for the Eco-Gourmet.* Boston: Beacon Press, 1977.

Wood, Rebecca. *The Whole Foods Encyclopedia: A Shopper's Guide.* New York: Prentice Hall Press, 1988.

Index

Italics indicate vegan recipes

Adzuki bean salad, Oriental, 185–186
Aloo gobi, 120–121
Amaranth cornbread, 234
Anadama bread, 217–218
Apple(s)
  brown rice with curried fruit, 150
  bulgur, zucchini, 146
  noodle pudding, fruity, 179
  pancakes, German, 253–254
  salad with walnut-leek dressing, green bean
    and endive, 210
  soup, lentil-, 71
Arugula, cannellini with fennel and, sautéed,
  110–111
Asian millet salad, 198
Asparagus salad with green pepper dressing,
  chickpea, 184–185
Asparagus and carrot soup, cream of, 94–95
Autumn quinoa and butter beans, 104

Baked chickpeas with eggplant and tomatoes,
  103
Baked ziti and eggplant with basil-tomato
  sauce, 136
Bangkok noodles, 123–124
Barbecued beans, 169
Barley
  and beans with spinach, mushrooms, and
    pignoli, 102
  and Brussels, 159–160
  chowder, vegetable, 90–91
  harirah, 88–89
  onion, 160
  pilaf, 160–161
  soup, Canadian pea, 81
  soup, mushroom-, 82
  -stuffed green peppers, 105
Bean(s). *See also* Black; Black-eyed peas; Butter; Cannellini; Chickpea; Kidney; Lentil;
  Lima; Pinto; Split pea; Tofu; White
  baked, Boston, 168–169
  barbecued, 169
  barley and, with spinach, mushrooms, and
    pignoli, 102
  canned beans, 50–51
  cooking
    age of beans, 46
    altitude, 47
    charts

    canned bean equivalents, 50, 70
    pressure cooker timetable, 57
    stovetop timetable, 53
  cookware used, 45–46
    pressure cooking, 56–57
    salted water, 49
    soaking, 48–49
    symbols, 67
    testing for doneness, 49–50
    water softness, 47
    water temperature, 45
  gas, 47–48
  interchangeable beans, 52–53
  mail-order sources, 63–64
  serving sizes, 67–68
  fava, and bulgur, Mediterranean, 101
  pasta and, with pesto, 134
  refried, 170
  salad
    adzuki, Oriental, 185–186
    asparagus and chickpea, with green pep-
      percorn dressing, 184–185
    black-eyed pea, wild rice, yellow pepper
      and, 180
    broccoli-, with balsamic dressing, 184
    hearts of palm and chickpea, 181
    hoppin' John, 193
    kidney, with artichokes and olives, red
      and white, 182–183
    kidney, with avocado dressing, 188–189
    lentil-bulgur, with feta cheese, 191
    lentil-potato, 189
    marinated chickpea, 183
    Mexicorn-lentil, 191–192
    millet-lentil, with cinnamon-orange dress-
      ing, 194–195
    millet and, with coriander-lime dressing,
      192
    onion, rice and, 193–194
    pinto with honey-ginger dressing, 181–
      182
    three-, traditional, 187–188
    tomato, Vidalia onion, and chickpea, 187
    wheat berry, orange and, 190
    white, –cauliflower, Spanish, 186
  soup
    three-, 75–76
    minestrone, 83–84
    and parsnip, pureed white, 97–98
    seven-, 76–77
    vegetable-, 74–75

Bean(s) (cont.)
    spread, Spanish-style, 247–248
    sprouts
        chow mein, 129
        lo mein, 130
        rice, fried, 155
        salad with calypso dressing, tropical, 199–
            200
        salad, with grapefruit-mint dressing, and
            watercress, 200–201
        tofu with cloud ears, spicy, 127
        tostadas, 141–142
Biscuits, buttermilk, 233–234
Black bean(s)
    chili, quick vegetable, 117
    chili, red, red, 116
    sauce, polenta with eggplant and, 139–140
    soup, Madeira–, 173
    soup, seven-bean, 76–77
Black-eyed peas
    salad, hoppin' John, 193
    salad, wild rice, yellow pepper, and, 180
    soup, seven-bean, 76–77
Blintzes, cheese, 260–261
Boston baked beans, 168–169
Bran
    bread, molasses-oatmeal, 214–215
    cereal, crazy mixed-up, 264
    granola, 262
    muesli, 263
    muffins, molasses–raisin, 238
    muffins, zucchini-banana, 239
    oatmeal, extra-high-fiber, 263–264
    pancakes, cinnamon–cottage cheese, 255
    pancakes with strawberry-orange sauce, oat,
        252–253
    waffles, pumpkin pie, 257
Braised French lentils, 167–168
Bread(s). See also Muffins
    quick
        amaranth corn-, 234
        biscuits, buttermilk, 233–234
        cranberry-pumpkin, 236
        honey-orange, 237
        teff banana, 235
    yeast
        anadama, 217–218
        garlic bites, 229–230
        graham, 215–216
        molasses-oatmeal, 214–215
        pepper-pear, 230–231
        pizza, red pepper, 227–229
        pumpernickel, 222–223
        rolls, walnut-raisin, 232–233
        rye, light, 218–219
        rye, corn, 220–221
        three-grain, 221–222
        wheat berry, 225–226
        whole wheat baguette, 224–225
Breakfast dishes
    blintzes, cheese, 260–261
    cereal
        crazy mixed-up, 264
        granola, 262
        muesli, 263

    oatmeal, extra-high-fiber, 263–264
    pancakes
        cinnamon–cottage cheese, 255
        German apple, 253–254
        honey-whole wheat, 251–252
        oat bran with strawberry-orange sauce,
            252–253
    spoonbread, orange, 261–262
    waffles
        cornmeal, 258
        orange-graham, 256
        pumpkin pie, 257
Broccoflower salad, curried, 104
Broccoli
    with garlic sauce, 126
    pasta with, and zucchini in garlic broth, 132
    penne from heaven, 131
    salad with balsamic dressing, -bean, 184
    soup, house-special, 87–88
Broth, 58–61
Brown rice with curried fruit, 140
Brussels and barley, 159–160
Buckwheat. See Kasha
Bulgur
    fava beans and, Mediterranean, 101
    salad
        with feta cheese, lentil-, 191
        with buttermilk dressing, vegetable, 196–
            197
        tabouli, 197
    zucchini-apple, 146
Butter bean(s)
    quinoa and, autumn, 104
    ratatouille, and brown rice, 107
    squash, greens, and, honey-glazed, 171–172
    stuffed acorn squash with, and chestnuts,
        170–171
Buttermilk biscuits, 233–234

Cabbage
    and kasha and bow ties, 165–166
    chana with rice pilau, curried, 121–122
    chickpeas and vegetables Madras, 118–119
    chowder, vegetable, 90–91
    kasha, and beans, 108
    soup
        house-special, 87–88
        minestrone, 83–84
        seven-bean, 76–77
        shredded, 85
    stew with millet, Senegal, 108–109
Cajun lentil stew, 114
Canadian pea soup, 81
Cannellini
    with fennel and sautéed arugula, 110–111
    soup, Swiss chard, 87
    salad
        with balsamic dressing, broccoli-bean, 184
        millet and bean, with coriander-lime dress-
            ing, 192–193
        wheat berry, orange, 190
Carrot(y)
    potage Crecy, 95–96
    puree of squash and, orange, 177
    quinoa, 163

salsify and, sherried, 178
soup, cream of asparagus and, 94–95
Cauliflower
  aloo gobi, 120–121
  chickpeas and vegetables Madras, 118–119
  lentils, braised French, 167–168
  salad, Spanish white bean, 186
Cereal
  crazy mixed-up, 264
  granola, 262
  muesli, 263
  oatmeal, extra-high-fiber, 263–264
Charts
  bean pressure cooker timetable, 57
  bean stovetop timetable, 53
  calorie intake table, 18
  canned bean equivalents, 50, 70
  fat comparison table, 13
  grain microwaving timetable, 55
  grain pressure cooker, natural release, 58
  grain pressure cooker, quick release, 58
  grain stovetop timetable, 54
  protein comparison table, 11
  sodium levels in common foods, 28–29
  vegetable shopping guide, 68–70
Cheese blintzes, 260
Chickpea(s)
  with aromatic rice, herbed, 109–110
  baked with eggplant and tomatoes, 103
  curried chana with rice pilau, 121–122
  quinoa-sunchoke pilaf, 161–162
  salad
    asparagus and, with green peppercorn
      dressing, 184–185
    hearts of palm and, 181
    marinated, 183
    three-bean, traditional, 187–188
    tomato, Vidalia onion and, 187
  stew with couscous, Moroccan, 100–101
  stew with millet, Senegal, 108–109
  stewed, and okra, 166
  and vegetables Madras, 118–119
Chili
  red, red, 116
  potato wedges, 172–173
  vegetable, quick, 117
  white bean, with wheat berries, 115
Chopped liver, mock, 248
Chow mein, 129
Chunky salad with gazpacho dressing, 202
Cinnamon–cottage cheese pancakes, 255
Corn. See also Cornmeal; Polenta
  rice, fried, 155
  salad, marinated chickpea, 183
  salad, Mexicorn-lentil, 191–192
  soup
    chowder, 91–92
    chowder, vegetable, 90–91
    cream of, with red pepper puree, 93–94
    fresh vegetable, 88–89
    succotash, Shaker, 92–93
Cornmeal
  bread
    Amaranth corn-, 234
    anadama, 217–218

corn rye, 220–221
  muffins, 240
  pumpernickel, 222–223
  three-grain, 221–222
  spoonbread, orange, 261–262
  pizza, Mexican, 142–145
  waffles, 258
Couscous
  salad with calypso dressing, tropical, 199–
    200
  stew with, Moroccan, 100–101
  with three peppers, 147
Cranberry-pumpkin bread, 236
Cream of asparagus and carrot soup, 94–95
Cream of corn soup with red pepper puree,
  93–94
Crepes, whole wheat, 259–260
Cucumber rice, 106
Cucumber salad, wilted, 206–207
Curry(ied)
  aloo gobi, 120–121
  brown rice with, fruit, 150
  chana with rice pilau, 121–122
  chickpeas and vegetables Madras, 118–119
  salad with calypso dressing, tropical, 199–
    200
  salad, tofu, quinoa, and hearts of palm,
    198–199
  soup, squash, 96–97
  spinach, tofu, peas, and potatoes, 117–118
  stew with millet, Senegal, 108–109

Dal, 79–80
Diced vegetable salad, 202–203
Dilled lima beans, 167
Dressing(s)
  avocado, 188–189
  avocado and blue cheese, 206
  balsamic, 184
  buttermilk, 196–197
  calypso, 199–200
  cinnamon orange, 194–195
  coriander-lime, 192
  gazpacho, 202
  grapefruit-mint, 200–201
  guilt-free
    garlic-pepper, 210
    herb, 209–210
    Roquefort, 208
    Thousand Island, 209
  honey-ginger, 181–182
  peppercorn, green, 184–185
  tahini, 211
  walnut-leek, 201

Eating Right Pyramid, 5
Eggplant
  chickpeas with, and tomatoes, baked, 103
  fava beans and bulgur, Mediterranean, 101
  with fresh figs, fragrant, 124–125
  lasagne, vegetable, 137–138
  pie, ratatouille, 111–112
  polenta with, and black bean sauce, 139–
    140
  ratatouille, butter beans, and brown rice, 107

Eggplant (cont.)
    ziti and eggplant with basil-tomato sauce,
        baked, 136–137

Fennel and sautéed arugula, cannellini, with,
    110–111
Food plan
    eating plan, 22–24
    meal planning, 32–34
    menus, 35–41
    Modified eating plan, 24
    strategies for the lacto-ovo vegetarian, 31–
        32
    strategies for the Modified plan, 29–31
Fragrant eggplant with fresh figs, 124–125
Fresh vegetable soup, 88–89
Fried rice, 155
Fruity noodle pudding, 179

Garlic bites, 229–230
Gazpacho, 89
German apple pancakes, 253–254
Graham bread, 215–216
Granola, 262
Grain(s). See also Barley; Bran; Bread; Bulgur;
    Cereal; Cornmeal; Couscous; Kasha; Mil-
    let; Oat; Polenta; Quinoa; Rice; Wheat
    berries; Whole wheat; Wild rice
    cooking
        age of grain, 46
        altitude, 47
        charts
            microwaving timetable, 55
            pressure cooker, natural release, 57
            pressure cooker, quick release, 58
            stovetop timetable, 54
        cookware used, 45–46
        pressure cooking, 56–58
        serving sizes, 70–71
        symbols, 70
        testing for doneness, 49–50
        water softness, 47
        water temperature, 45
        interchangeable grains, 51
        mail-order sources, 63–64
Green beans
    salad
        and endive, with walnut-leek dressing, 201
        –kohlrabi, 205
        three-bean, traditional, 187–188
    soup, green lentil and vegetable, 78–79
    soup, vegetable, fresh, 84–85
    wheat berries with celery and, 147–148
Green gazpacho, 91
Green lentil and vegetable soup, 78
Grilled vegetables, Oriental, 175–176
Guilt-free dressing
    herb, 209–210
    garlic-pepper, 210
    Roquefort, 208
    Thousand Island, 209

Harirah, 88–89
Hearts of palm and chickpea salad, 181
Herb and garlic creamy cheese, 243

Herbed chickpeas with aromatic rice, 109–
    110
Herbed scalloped potatoes, 173–174
Honey
    bread, -orange, 237
    pancakes, –whole wheat, 251
    squash, greens, and beans, -glazed, 171–
        172
Hoppin' John salad, 193
House-special soup, 87–88
Hummos, 247

Italian lentil-vegetable soup, 77–78

Jambalaya rice and beans, 99

Kasha
    and bow ties, 165–166
    cabbage, and beans, 108
    and cauliflower, 165
Kidney bean(s)
    barbecued beans, 169
    chili, red, red, 116
    rice and, jambalaya, 99
    salad
        with artichokes and olives, red and white,
            182–183
        with avocado dressing, 188–189
        pasta, yuppie, 195–196
        three-bean, traditional, 187–188
    soup, seven-bean, 76–77

Lasagne, vegetable, 137–138
Lemon-herb quinoa, 164
Lentil(s)
    braised French, 167–168
    red, and cucumber rice, 106
    salad
        -bulgur with feta cheese, 191
        potato
        Mexicorn, 191–192
        millet-, with cinnamon-orange dressing,
            194–195
    soup
        -apple, 71
        dal, 79–80
        harirah, 88–89
        Italian -vegetable, 77–78
        sweet potato–red, 73
        vegetable-bean, 74–75
    stew, Cajun, 114
Light rye bread, 218–219
Lima bean(s)
    cabbage, kasha, and, 108
    dilled, 167
    soup
        chowder, vegetable, 90–91
        three-bean, 75–76
        seven-bean, 75–76
        vegetable bean, 74–75
Liptauer, 244
Lo mein, 130

Madeira–black bean soup, 72
Marinated chickpea salad, 183

Mediterranean fava beans and bulgur, 101
Muesli, 263
Mexican pizza, 142–145
Millet
  Asian, salad, 198
  salad with coriander-lime dressing, and
    bean, 193
  salad with cinnamon-orange dressing, -lentil,
    194–195
Mexicorn-lentil salad, 191–192
Midsummer's pasta, 135
Minestrone, 83
Mock chopped liver, 248
Molasses oatmeal bread, 214
Molasses–raisin bran muffins, 238
Moroccan stew with couscous, 100–101
Muffins
  corn, 140
  –raisin bran, molasses, 238
  whole wheat, 241
  zucchini-banana bran, 239
Mushrooms
  barley and beans with spinach, and pignoli,
    102
  chopped liver, mock, 248
  lo mein, 130
  polenta, porcini, 158–159
  rice pilaf, 151–152
  soup
    -barley, 82
    lentil and vegetable, green, 78
    vegetable-bean, 74
  squash with vegetable sauce, spaghetti, 113–
    114
  stuffed green peppers, barley-, 105
  wheat berries with oyster, 148–149

Naked tomato salad, 203
New Four Food Groups Food plan
  eating plan, 22–24
  meal planning, 32–34
  menus, 35–41
  Modified eating plan, 24
  strategies for the lacto-ovo vegetarian, 31–
    32
  strategies for the Modified plan, 29–31
Noodle(s). *See also* Pasta
  Bangkok, 123–124
  harirah, 88–89
  lo mein, 130
  with peanut sauce, 122–123
  pudding, fruity, 179
Nutritional information
  calcium, 20–21
  calories, 19
  carbohydrates, 11–12, 29
  cholesterol, 15–16
  Eating Right Pyramid, 5
  fats, 13–19
  fiber, 20
  iron, 21
  "old" four food groups, 3, 4
  protein primer, 6–11
  sodium watching, 27–29
  strategies, 26–29

  weight control, 25
  vitamin D, 21
  vitamin B_{12}, 21

Oat(s)
  bread, molasses-, 214–215
  cereal, crazy mixed-up, 264
  granola, 262
  -meal, extra-high-fiber, 263–264
  muesli, 263
  pancakes with strawberry-orange sauce,
    bran, 252–253
Old-fashioned split pea soup, 80–81
Olive creamy cheese, 242
Onion
  barley, 160
  salad, rice, and bean, 193–194
  soup, 86
Orange
  spoonbread, 261–262
  squash and carrots, puree of, 177
  waffles, graham, 256
Oriental adzuki bean salad, 185–186
Oriental grilled vegetables, 175–176

Pancakes
  cinnamon cottage cheese, 255
  German apple, 253–254
  honey–whole wheat, 251–252
  oat bran with strawberry-orange sauce,
    252–253
Pasta. *See also* Noodle(s)
  and beans with pesto, 134
  with broccoli and zucchini in garlic broth,
    132
  kasha and bow ties, 165–166
  lasagne, vegetable, 137–138
  midsummer's, 135
  minestrone, 83–84
  penne from heaven, 131
  salad, yuppie, 195–196
  ziti and eggplant with basil-tomato sauce,
    baked, 136–137
  ziti with white bean marinara sauce, 133
Pea(s). *See also* Black-eyed; Chick-; Split
  aloo gobi, 120–121
  barley, onion, 160
  chickpeas and vegetables Madras, 118–119
  curried chana with rice pilau, 121–122
  curried spinach, tofu, and potatoes, 117–
    118
  lo mein, 130
  quinoa-sunchoke pilaf, 161–162
  rice, fried, 155
  rice pilaf, 151–152
  rice pilaf, white and wild, 156
  salad
    millet, Asian, 198
    pea, 204–205
    tofu, quinoa, and hearts of palm, 198–
      199
  soup, house-special, 87–88
Peanut butter, honey-orange, 246
Peanut sauce, noodles with, 122–123
Penne from heaven, 131

Pepper-pear bread, 230–231
Pesto, pasta and beans with, 134
Pilaf
    barley, 160–161
    quinoa-sunchoke, 161–162
    rice, 151–152
    rice, white and wild, 156
Pilau, curried chana with rice, 121–122
Pinto bean(s)
    pizza, Mexican, 142–145
    refried, 170
    salad with honey-ginger dressing, 181–182
    soup, seven-bean, 76–77
Pizza bread, red pepper, 227–229
Pizza, Mexican, 142–145
Polenta
    with eggplant and black bean sauce, 139–
        140
    pie, cheese, 140–141
    pie, ratatouille, 111–112
    porcini, 158–159
Porcini polenta, 158–159
Potage Crecy, 95–96
Potato(es). See also Sweet potato(es)
    aloo gobi, 120–121
    curried chana with rice pilau, 121–122
    curried spinach, tofu, pea, and, 117–118
    potage Crecy, 95–96
    salad, lentil-, 189
    scalloped, herbed, 173–174
    stuffed baked, 174–175
    wedges, chili, 172–173
Provençale spread, 249–250
Pumpernickel, 222–223
Pumpkin pie waffles, 257
Pureed white bean and parsnip soup, 97–98

Quick vegetable chili, 117
Quinoa
    and butter beans, autumn, 104
    carroty, 163
    lemon-herb, 164
    pilaf, -sunchoke, 161–162
    salad, tofu, and hearts of palm, 198–199
    -stuffed tomatoes, 162–163

Ratatouille, butter beans, and brown rice, 107
Ratatouille pie, 111–112
Red
    chili, red, 116
    lentils and cucumber rice, 106
    pepper pizza bread, 227–229
    pepper puree, 93–94
    salad with artichokes and olives, and white
        kidney beans, 182–183
Refried beans, 170
Rice. See also Wild rice
    brown
        with curried fruit, 150
        ratatouille, butter beans, and, 107
        salad, hoppin' John, 193
    risotto with fresh herbs, 152–153
    risotto, zucchini and yellow pepper, 153–
        154

white
    and beans, jambalaya, 99
    chickpeas with aromatic, herbed, 109–110
    with East Indian flavors, 150–151
    fried, 155
    lentils and cucumber, red, 106
    pilaf, 151–152
    pilaf, white and wild, 156
    pilau, curried chana with, 121–122
    salad, onion, and bean, 193–194
Risotto with fresh herbs, 152–153
Risotto, zucchini and yellow pepper, 153–154
Rye bread
    corn, 220–221
    light, 218–219
    pumpernickel, 222–223
    three-grain, 221–222
Salad(s)
    adzuki bean, Oriental, 185–186
    asparagus and chickpea, with green pepper-
        corn dressing, 184–185
    bean sprout and watercress, with grapefruit-
        mint dressing, 200–201
    broccoli-bean with balsamic dressing, 184
    broccoflower, curried, 204
    chickpea, marinated, 183
    chunky, with gazpacho dressing, 202
    cucumber, wilted, 206–207
    dressing
        guilt-free
            garlic-pepper, 210
            herb, 209–210
            Roquefort, 208
            Thousand Island, 209
        tahini, 211
    green beans and endive, with walnut-leek
        dressing, 201
    green bean-kohlrabi, 205
    hearts of palm and chickpea, 181
    hoppin' John, 193
    kidney bean, with artichokes and olives, red
        and white, 182–183
    kidney bean, with avocado dressing, 188–
        189
    lentil-bulgur, with feta cheese, 191
    lentil-potato, 189
    pasta, yuppie, 195–196
    pea, 204–205
    Mexicorn-lentil, 191–192
    millet
        Asian, 198
        and bean, with coriander-lime dressing,
            192
        -lentil, with cinnamon-orange dressing,
            194–195
    onion, rice, and bean, 193–194
    pinto with honey-ginger dressing, 181–182
    tabouli, 197
    three-bean, traditional, 187–188
    tofu, quinoa, and hearts of palm, 198–199
    tomato, naked, 203
    tomato, Vidalia onion, and chickpea, 187
    tossed, with avocado and blue cheese dress-
        ing, 206

tropical, with calypso dressing, 199–200
vegetable-bulgur, with buttermilk dressing,
196–197
vegetable, diced, 202–203
wheat berry, orange, and bean, 190
white bean–cauliflower, Spanish, 186
wild rice, yellow pepper, and black-eyed pea,
180
Sautéed baby squash and fennel, 176
Scalloped potatoes, herbed, 173–174
Senegal stew with millet, 108–109
Serving sizes, 67
Sesame spread, 244–245
Sesame tofu and vegetables, 128
Seven-bean soup, 76–77
Shaker succotash soup, 92–93
Sherried salsify and carrots, 178
Shredded cabbage soup, 85
Soup(s)
  cabbage, shredded, 85
  Canadian pea, 81
  corn chowder, 91–92
  cream of asparagus and carrot, 94–95
  cream of corn with red pepper puree, 93–94
  curried squash, 96–97
  dal, 79–80
  gazpacho, 89
  green gazpacho, 90
  green lentil and vegetable, 78
  harirah, 88–89
  house-special, 87–88
  Italian lentil-vegetable, 77
  lentil-apple, 71
  Madeira–black bean, 72
  minestrone, 83–84
  mushroom-barley, 82
  onion, 86
  potage Crecy, 95–96
  seven-bean, 76–77
  Shaker succotash, 92–93
  split pea, old-fashioned, 80–81
  sweet potato–red lentil, 73
  Swiss chard, 87
  three-bean, 75–76
  vegetable
    -bean, 74–75
    chowder, 90–91
    fresh, 84–85
  white bean and parsnip, pureed, 97–98
Spaghetti squash with vegetable sauce, 113–
  114
Spanish-style bean spread, 247–248
Spanish white bean–cauliflower salad, 186
Spicy tofu with cloud ears, 127
Spinach
  barley and beans with, mushrooms, and pig-
    noli, 102
  curried, tofu, peas, and potatoes, 117–118
  salad, hoppin' John, 193
Split pea(s)
  soup
    Canadian, 81
    harirah, 88–89
    old-fashioned, 80–81

seven-bean, 76–77
Spoonbread, orange, 261–262
Spreads
  bean, Spanish-style, 247–248
  chopped liver, mock, 248
  herb and garlic creamy cheese, 243
  hummos, 247
  Liptauer, 244
  olive creamy cheese, 242
  Provençale, 249–250
  peanut butter, honey-orange, 246
  sesame, 244–245
  tofu fruit cheese, 245–246
Squash. *See also* Zucchini
  summer
    Oriental grilled vegetables, 175–176
    puree of, and carrots, orange, 177
    sautéed baby, and fennel, 176
    soup, Italian lentil-vegetable, 77
  winter
    acorn with butter beans and chestnuts,
      stuffed, 170–171
    honey-glazed, greens, and beans, 171–172
    quinoa and butter beans, autumn, 104
    soup, curried, 96–97
    spaghetti, with vegetable sauce, 113–114
    stew with couscous, Moroccan, 100–101
Stew(ed)
  chickpeas and okra, 166
  Moroccan, with couscous, 100–101
  Senegal, with millet, 108–109
Stocking the pantry, 60–62
Stuffed
  acorn squash with butter beans and chest-
    nuts, 170
  green peppers, barley-, 105
  potatoes, baked, 174–175
Sweet potato(es)
  quinoa and butter beans, autumn, 104
  soup, –red lentil, 73
  stew with couscous, Moroccan, 100–101
  stew with millet, Senegal, 108–109
Swiss chard soup, 87
Symbols in recipe titles, 67

Tabouli, 197
Tahini dressing, 211
Teff banana bread, 235
Three-bean soup, 75–76
Three-grain bread, 221–222
Traditional three-bean salad, 187
Tofu
  Bangkok noodles, 123–124
  broccoli with garlic sauce, 126
  with cloud ears, spicy, 127
  curried spinach, peas, and potatoes, 117–
    118
  fruit cheese, 245–246
  salad, Asian millet, 198
  salad, quinoa, and hearts of palm, 198–199
  soup, house-special, 87–88
  and vegetables, sesame, 128
Tomato(es)
  chickpeas with eggplant and, baked, 103

Tomato(es) (cont.)
marinara sauce, ziti with white bean, 133
salad, naked, 203
salad, Vidalia onion and chickpea, 187
stuffed, quinoa, 162–163
Tossed salad with avocado and blue cheese dressing, 206
Tostadas, 141–142
Tropical salad with calypso dressing, 199–200

Vegetable(s). See also Broccoli; Cabbage; Carrot; Cauliflower; Corn; Eggplant; Green beans; Mushrooms; Onion; Peas; Potato; Spinach; Squash; Sweet potato(es); Tomato; Zucchini
chickpeas and, Madras, 118–119
chili, quick, 117
cucumber rice, 106
cucumber salad, wilted, 206–207
grilled, Oriental, 175–176
lasagne, 137–138
peppers, barley-stuffed green, 105
salad, with buttermilk dressing, -bulgur, 196–197
salad, diced, 202–203
sauce, spaghetti squash with, 113–114
sautéed arugula, cannellini with fennel and, 110–111
sesame tofu and, 128
soup
-bean, 74–75
chowder, 90–91
fresh, 84–85
green lentil and, 78
Italian lentil, 77–78
shopping guide, 68–70
stocking the pantry, 61
wheat berries and Oriental, 149

Waffles
cornmeal, 258
orange-graham, 256
pumpkin pie, 257
Walnut-raisin rolls, 232–233
Wheat berry(ies). See also Whole wheat bread, 225–226
with celery and green beans, 147–148
chili, white bean, 115
and Oriental vegetables, 149
with oyster mushrooms, 148–149
salad, orange and bean, 190

White bean(s)
barbecued, 169
Boston baked, 168–169
chili, with wheat berries, 115
salad, Spanish –cauliflower, 186
soup, pureed, and parsnip, 97–98
soup, seven-bean, 76–77
ziti with, marinara sauce, 133
White and wild rice pilaf, 156
Whole wheat
baguette, 224–225
crepes, 259
muffins, 241
pancakes, honey, 251
Wild rice
pilaf, white and, 156
with sugar snaps, 157–158
yellow pepper, and black-eyed pea salad, 180
zucchini, leeks and, 157
Wilted cucumber salad, 206–207

Yeast, baking with, 212–214
Yuppie pasta salad, 195–196

Ziti with white bean marinara sauce, 133
Zucchini
bulgur, apple, 146
lasagne, vegetable, 137–138
muffins, -banana bran, 239
pasta with broccoli and, in garlic broth, 132
penne from heaven, 131
pie, ratatouille, 111–112
ratatouille, butter beans, and brown rice, 107
rice pilaf, 151–152
risotto, and yellow pepper, 153–154
salad
broccoli-bean with balsamic dressing, 184
pasta, yuppie, 195–196
vegetable-bulgur, with buttermilk dressing, 196–197
wild rice, yellow pepper, and black-eyed pea, 180
soup
Italian lentil-vegetable, 77
minestrone, 83–84
vegetable, fresh, 84–85
stew, Cajun lentil, 114
wild rice, leeks and, 157